Kai-Philipp Walter Pahl

Distributed Transformers for Broadband Monolithic Millimeter-Wave Integrated Power Amplifiers

Karlsruher Forschungsberichte
aus dem Institut für Hochfrequenztechnik und Elektronik

Herausgeber: Prof. Dr.-Ing. Thomas Zwick

Band 80

Distributed Transformers for Broadband Monolithic Millimeter-Wave Integrated Power Amplifiers

by
Kai-Philipp Walter Pahl

KIT Scientific Publishing

Dissertation, Karlsruher Institut für Technologie (KIT)
Fakultät für Elektrotechnik und Informationstechnik, 2014

Impressum

KIT Scientific Publishing

Karlsruher Institut für Technologie (KIT)
KIT Scientific Publishing
Straße am Forum 2
D-76131 Karlsruhe

KIT Scientific Publishing is a registered trademark of Karlsruhe Institute of Technology. Reprint using the book cover is not allowed.

www.ksp.kit.edu

Print on Demand 2015

ISSN 1868-4696
ISBN 978-3-7315-0409-2
DOI 10.5445/KSP/1000048002

Editor's Preface

The enormous technological progress accomplished over the last decades facilitates the utilization of millimeter-wave frequencies for mass products like auto motive radar, industrial sensors, high-speed data communication links or medical applications. Besides novel and cost-efficient packaging technologies, in particular semiconductor technologies (e.g. SiGe, CMOS, GaAs, GaN, InP, etc.) with constantly improving cut-off frequencies are essential for the realization of cost-efficient and compact millimeter-wave modules. The dominant limitation factor of all semiconductor technologies is the available output power. The realization of faster semiconductor devices requires down scaling, which on the other hand results in a reduced breakdown voltage and finally less output power. While millimeter-wave amplifiers based on GaN achieve an output power of about 1 W at 100 GHz, above 200 GHz barely more than 100 mW have been achieved using InP technologies. Another issue is that amplifiers optimized for maximum output power usually only provide a limited bandwidth, which often is not sufficient for many communication, radar and also measurement applications.

The key issue of having output power in conjunction with bandwidth is addressed in Mr. Kai-Philipp Walter Pahl's thesis. To design an amplifier that provides more output power than that of a single transistor device, multiple transistors are put in parallel. However, the maximum number of parallel devices that actually leads to an increased overall output power is limited, since the loss of the distribution network increases more than linearly with the number of transistors. The reasons for this are the increasing transmission lines lengths, which are necessary to separate the active devices and to provide the required impedance transformation. The higher the number of parallel transistors the higher the impedance ratio between the parallel transistors and the single common port of the amplifier. In addition a high transformation ratio results in a reduced bandwidth. One way to solve this issue is to use magnetic transformers for simultaneous power distribution and impedance matching. A group around Hajimiri and Rutledge first proposed such an approach in 2001. In his dissertation Mr. Pahl has developed important scientific fundamentals for the development of monolithic integrated broadband amplifiers based on distributed transformers. By using a novel geometry of multiple distributed transformers, optimized for an application at millimeter-wave frequencies, he was able to demonstrate superior performance compared to the standard circular geometry known from literature. In particular the significantly improved design equations for efficiency optimized distributed transformers, which are based on an improved equivalent circuit model, but also novel design equations for distributed transformers with maximum symmetrical bandwidth, were a huge contribution to the scientific field and extended the state of the art significantly. I am absolutely sure that this work will attract much interest from the research community worldwide with many circuit realizations based on. My personal wish for Mr. Pahl is that his creativity and great innovative capacity will continue to earn him both scientific and economic success.

Prof. Dr.-Ing. Thomas Zwick
– Director of the IHE –

Forschungsberichte aus dem Institut für Höchstfrequenztechnik und Elektronik (IHE) der Universität Karlsruhe (TH) (ISSN 0942-2935)

Herausgeber: Prof. Dr.-Ing. Dr. h.c. Dr.-Ing. E.h. mult. Werner Wiesbeck

Band 1 Daniel Kähny
Modellierung und meßtechnische Verifikation polarimetrischer, mono- und bistatischer Radarsignaturen und deren Klassifizierung (1992)

Band 2 Eberhardt Heidrich
Theoretische und experimentelle Charakterisierung der polarimetrischen Strahlungs- und Streueigenschaften von Antennen (1992)

Band 3 Thomas Kürner
Charakterisierung digitaler Funksysteme mit einem breitbandigen Wellenausbreitungsmodell (1993)

Band 4 Jürgen Kehrbeck
Mikrowellen-Doppler-Sensor zur Geschwindigkeits- und Wegmessung - System-Modellierung und Verifikation (1993)

Band 5 Christian Bornkessel
Analyse und Optimierung der elektrodynamischen Eigenschaften von EMV-Absorberkammern durch numerische Feldberechnung (1994)

Band 6 Rainer Speck
Hochempfindliche Impedanzmessungen an Supraleiter / Festelektrolyt-Kontakten (1994)

Band 7 Edward Pillai
Derivation of Equivalent Circuits for Multilayer PCB and Chip Package Discontinuities Using Full Wave Models (1995)

Band 8 Dieter J. Cichon
Strahlenoptische Modellierung der Wellenausbreitung in urbanen Mikro- und Pikofunkzellen (1994)

Band 9 Gerd Gottwald
Numerische Analyse konformer Streifenleitungsantennen in mehrlagigen Zylindern mittels der Spektralbereichsmethode (1995)

Band 10 Norbert Geng
Modellierung der Ausbreitung elektromagnetischer Wellen in Funksystemen durch Lösung der parabolischen Approximation der Helmholtz-Gleichung (1996)

Band 11 Torsten C. Becker
Verfahren und Kriterien zur Planung von Gleichwellennetzen für den Digitalen Hörrundfunk DAB (Digital Audio Broadcasting) (1996)

Forschungsberichte aus dem
Institut für Höchstfrequenztechnik und Elektronik (IHE)
der Universität Karlsruhe (TH) (ISSN 0942-2935)

Band 12 Friedhelm Rostan
Dual polarisierte Microstrip-Patch-Arrays für zukünftige satellitengestützte SAR-Systeme (1996)

Band 13 Markus Demmler
Vektorkorrigiertes Großsignal-Meßsystem zur nichtlinearen Charakterisierung von Mikrowellentransistoren (1996)

Band 14 Andreas Froese
Elektrochemisches Phasengrenzverhalten von Supraleitern (1996)

Band 15 Jürgen v. Hagen
Wide Band Electromagnetic Aperture Coupling to a Cavity: An Integral Representation Based Model (1997)

Band 16 Ralf Pötzschke
Nanostrukturierung von Festkörperflächen durch elektrochemische Metallphasenbildung (1998)

Band 17 Jean Parlebas
Numerische Berechnung mehrlagiger dualer planarer Antennen mit koplanarer Speisung (1998)

Band 18 Frank Demmerle
Bikonische Antenne mit mehrmodiger Anregung für den räumlichen Mehrfachzugriff (SDMA) (1998)

Band 19 Eckard Steiger
Modellierung der Ausbreitung in extrakorporalen Therapien eingesetzter Ultraschallimpulse hoher Intensität (1998)

Band 20 Frederik Küchen
Auf Wellenausbreitungsmodellen basierende Planung terrestrischer COFDM-Gleichwellennetze für den mobilen Empfang (1998)

Band 21 Klaus Schmitt
Dreidimensionale, interferometrische Radarverfahren im Nahbereich und ihre meßtechnische Verifikation (1998)

Band 22 Frederik Küchen, Torsten C. Becker, Werner Wiesbeck
Grundlagen und Anwendungen von Planungswerkzeugen für den digitalen terrestrischen Rundfunk (1999)

Band 23 Thomas Zwick
Die Modellierung von richtungsaufgelösten Mehrwegegebäudefunkkanälen durch markierte Poisson-Prozesse (2000)

Forschungsberichte aus dem
Institut für Höchstfrequenztechnik und Elektronik (IHE)
der Universität Karlsruhe (TH) (ISSN 0942-2935)

Band 24 Dirk Didascalou
Ray-Optical Wave Propagation Modelling in Arbitrarily Shaped Tunnels (2000)

Band 25 Hans Rudolf
Increase of Information by Polarimetric Radar Systems (2000)

Band 26 Martin Döttling
Strahlenoptisches Wellenausbreitungsmodell und Systemstudien für den Satellitenmobilfunk (2000)

Band 27 Jens Haala
Analyse von Mikrowellenheizprozessen mittels selbstkonsistenter finiter Integrationsverfahren (2000)

Band 28 Eberhard Gschwendtner
Breitbandige Multifunktionsantennen für den konformen Einbau in Kraftfahrzeuge (2001)

Band 29 Dietmar Löffler
Breitbandige, zylinderkonforme Streifenleitungsantennen für den Einsatz in Kommunikation und Sensorik (2001)

Band 30 Xuemin Huang
Automatic Cell Planning for Mobile Network Design: Optimization Models and Algorithms (2001)

Band 31 Martin Fritzsche
Anwendung von Verfahren der Mustererkennung zur Detektion von Landminen mit Georadaren (2001)

Band 32 Siegfried Ginter
Selbstkonsistente Modellierung der Erhitzung von biologischem Gewebe durch hochintensiven Ultraschall (2002)

Band 33 Young Jin Park
Applications of Photonic Bandgap Structures with Arbitrary Surface Impedance to Luneburg Lenses for Automotive Radar (2002)

Band 34 Alexander Herschlein
Entwicklung numerischer Verfahren zur Feldberechnung konformer Antennen auf Oberflächen höherer Ordnung (2002)

Band 35 Ralph Schertlen
Mikrowellenprozessierung nanotechnologischer Strukturen am Beispiel von Zeolithen (2002)

Forschungsberichte aus dem
Institut für Höchstfrequenztechnik und Elektronik (IHE)
der Universität Karlsruhe (TH) (ISSN 0942-2935)

Band 36 Jürgen von Hagen
Numerical Algorithms for the Solution of Linear Systems of Equations Arising in Computational Electromagnetics (2002)

Band 37 Ying Zhang
Artificial Perfect Magnetic Conductor and its Application to Antennas (2003)

Band 38 Thomas M. Schäfer
Experimentelle und simulative Analyse der Funkwellenausbreitung in Kliniken (2003)

Band 39 Christian Fischer
Multistatisches Radar zur Lokalisierung von Objekten im Boden (2003)

Band 40 Yan C. Venot
Entwicklung und Integration eines Nahbereichsradarsensorsystems bei 76,5 GHz (2004)

Band 41 Christian Waldschmidt
Systemtheoretische und experimentelle Charakterisierung integrierbarer Antennenarrays (2004)

Band 42 Marwan Younis
Digital Beam-Forming for high Resolution Wide Swath Real and Synthetic Aperture Radar (2004)

Band 43 Jürgen Maurer
Strahlenoptisches Kanalmodell für die Fahrzeug-Fahrzeug-Funkkommunikation (2005)

Band 44 Florian Pivit
Multiband-Aperturantennen für Basisstationsanwendungen in rekonfigurierbaren Mobilfunksystemen (2005)

Band 45 Sergey Sevskiy
Multidirektionale logarithmisch-periodische Indoor-Basisstationsantennen (2006)

Band 46 Martin Fritz
Entwurf einer breitbandigen Leistungsendstufe für den Mobilfunk in Low Temperature Cofired Ceramic (2006)

Band 47 Christiane Kuhnert
Systemanalyse von Mehrantennen-Frontends (MIMO) (2006)

Band 48 Marco Liebler
Modellierung der dynamischen Wechselwirkungen von hoch-intensiven Ultraschallfeldern mit Kavitationsblasen (2006)

Forschungsberichte aus dem
Institut für Höchstfrequenztechnik und Elektronik (IHE)
der Universität Karlsruhe (TH) (ISSN 0942-2935)

Band 49 Thomas Dreyer
Systemmodellierung piezoelektrischer Sender zur Erzeugung hochintensiver Ultraschallimpulse für die medizinische Therapie (2006)

Band 50 Stephan Schulteis
Integration von Mehrantennensystemen in kleine mobile Geräte für multimediale Anwendungen (2007)

Band 51 Werner Sörgel
Charakterisierung von Antennen für die Ultra-Wideband-Technik (2007)

Band 52 Reiner Lenz
Hochpräzise, kalibrierte Transponder und Bodenempfänger für satellitengestützte SAR-Missionen (2007)

Band 53 Christoph Schwörer
Monolithisch integrierte HEMT-basierende Frequenzvervielfacher und Mischer oberhalb 100 GHz (2008)

Band 54 Karin Schuler
Intelligente Antennensysteme für Kraftfahrzeug-Nahbereichs-Radar-Sensorik (2007)

Band 55 Christian Römer
Slotted waveguide structures in phased array antennas (2008)

Fortführung als
"Karlsruher Forschungsberichte aus dem Institut für Hochfrequenztechnik und Elektronik" bei KIT Scientific Publishing
(ISSN 1868-4696)

Karlsruher Forschungsberichte aus dem
Institut für Hochfrequenztechnik und Elektronik
(ISSN 1868-4696)

Herausgeber: Prof. Dr.-Ing. Thomas Zwick

Die Bände sind unter www.ksp.kit.edu als PDF frei verfügbar oder als Druckausgabe bestellbar.

Band 55 Sandra Knörzer
Funkkanalmodellierung für OFDM-Kommunikationssysteme bei Hochgeschwindigkeitszügen (2009)
ISBN 978-3-86644-361-7

Band 56 Thomas Fügen
Richtungsaufgelöste Kanalmodellierung und Systemstudien für Mehrantennensysteme in urbanen Gebieten (2009)
ISBN 978-3-86644-420-1

Band 57 Elena Pancera
Strategies for Time Domain Characterization of UWB Components and Systems (2009)
ISBN 978-3-86644-417-1

Band 58 Jens Timmermann
Systemanalyse und Optimierung der Ultrabreitband-Übertragung (2010)
ISBN 978-3-86644-460-7

Band 59 Juan Pontes
Analysis and Design of Multiple Element Antennas for Urban Communication (2010)
ISBN 978-3-86644-513-0

Band 60 Andreas Lambrecht
True-Time-Delay Beamforming für ultrabreitbandige Systeme hoher Leistung (2010)
ISBN 978-3-86644-522-2

Band 61 Grzegorz Adamiuk
Methoden zur Realisierung von dual-orthogonal, linear polarisierten Antennen für die UWB-Technik (2010)
ISBN 978-3-86644-573-4

Band 62 Jutta Kühn
AlGaN/GaN-HEMT Power Amplifiers with Optimized Power-Added Efficiency for X-Band Applications (2011)
ISBN 978-3-86644-615-1

Karlsruher Forschungsberichte aus dem Institut für Hochfrequenztechnik und Elektronik (ISSN 1868-4696)

Band 63 Małgorzata Janson
Hybride Funkkanalmodellierung für ultrabreitbandige MIMO-Systeme (2011)
ISBN 978-3-86644-639-7

Band 64 Mario Pauli
Dekontaminierung verseuchter Böden durch Mikrowellenheizung (2011)
ISBN 978-3-86644-696-0

Band 65 Thorsten Kayser
Feldtheoretische Modellierung der Materialprozessierung mit Mikrowellen im Durchlaufbetrieb (2011)
ISBN 978-3-86644-719-6

Band 66 Christian Andreas Sturm
Gemeinsame Realisierung von Radar-Sensorik und Funkkommunikation mit OFDM-Signalen (2012)
ISBN 978-3-86644-879-7

Band 67 Huaming Wu
Motion Compensation for Near-Range Synthetic Aperture Radar Applications (2012)
ISBN 978-3-86644-906-0

Band 68 Friederike Brendel
Millimeter-Wave Radio-over-Fiber Links based on Mode-Locked Laser Diodes (2013)
ISBN 978-3-86644-986-2

Band 69 Lars Reichardt
Methodik für den Entwurf von kapazitätsoptimierten Mehrantennensystemen am Fahrzeug (2013)
ISBN 978-3-7315-0047-6

Band 70 Stefan Beer
Methoden und Techniken zur Integration von 122 GHz Antennen in miniaturisierte Radarsensoren (2013)
ISBN 978-3-7315-0051-3

Band 71 Łukasz Zwirełło
Realization Limits of Impulse-Radio UWB Indoor Localization Systems (2013)
ISBN 978-3-7315-0114-5

Band 72 Xuyang Li
Body Matched Antennas for Microwave Medical Applications (2014)
ISBN 978-3-7315-0147-3

Karlsruher Forschungsberichte aus dem
Institut für Hochfrequenztechnik und Elektronik
(ISSN 1868-4696)

Band 73 Sebastian Diebold
Transistor- und Leitungsmodellierung zum Entwurf von monolithisch integrierten Leistungsverstärkern für den hohen Millimeterwellen-Frequenzbereich (2014)
ISBN 978-3-7315-0161-9

Band 74 Christian Rusch
Integrierte, planare Leckwellenantennen für 3D-Millimeterwellen-Radarsysteme basierend auf dem holografischen Prinzip (2014)
ISBN 978-3-7315-0234-0

Band 75 Marlene Harter
Dreidimensional bildgebendes Radarsystem mit digitaler Strahlformung für industrielle Anwendungen (2014)
ISBN 978-3-7315-0249-4

Band 76 Michael A. Baldauf
Abhängigkeit der Exposition von der Zellgröße beim Mobilfunk unter Gewährleistung der Versorgung (2015)
ISBN 978-3-7315-0308-8

Band 77 Alicja Ossowska
Highly Resolved Synthetic Aperture Radar with Beam Steering (2015)
ISBN 978-3-7315-0315-6

Band 78 Małgorzata Dominika Brzeska
RF Modelling and Characterization of Tyre Pressure Sensors and Vehicle Access Systems (2015)
ISBN 978-3-7315-0348-4

Band 79 Ulrich Lewark
Aktive Frequenzvervielfacher zur Signalerzeugung im Millimeter- und Submillimeterwellen Frequenzbereich (2015)
ISBN 978-3-7315-0354-5

Band 80 Kai-Philipp Walter Pahl
Distributed Transformers for Broadband Monolithic Millimeter-Wave Integrated Power Amplifiers (2015)
ISBN 978-3-7315-0409-2

Distributed Transformers for Broadband Monolithic Millimeter-Wave Integrated Power Amplifiers

Zur Erlangung des akademischen Grades eines

DOKTOR-INGENIEURS

von der Fakultät für
Elektrotechnik und Informationstechnik
des Karlsruher Instituts für Technologie (KIT)

genehmigte

DISSERTATION

von

Dipl.-Ing. Kai-Philipp Walter Pahl

geb. in Mannheim

Tag der mündlichen Prüfung:	23. 09. 2014
Hauptreferent:	Prof. Dr.-Ing. Thomas Zwick
Korreferent:	Prof. Dr.-Ing. Ingmar Kallfass

Danksagung

Das Erarbeiten und Verfassen einer Dissertation lässt sich meiner Meinung nach ohne die kontinuierliche Unterstützung der Menschen um einen herum nur sehr schwer bewerkstelligen. Sei es durch Rückhalt im privaten Umfeld, fachliche Diskussionen mit Kollegen oder aber auch gezielte Ablenkung durch Freunde, haben sehr viele Menschen auf ganz unterschiedliche Art und Weise Anteil am Erfolg einer Promotion. Bei all diesen Menschen möchte ich mich bedanken.

Die vorliegende Dissertation entstand während meiner Zeit als wissenschaftlicher Mitarbeiter am Institut für Hochfrequenztechnik und Elektronik (IHE) des Karlsruher Instituts für Technologie (KIT). Die hervorragende Arbeitsatmosphäre und gute Stimmung am IHE, zu der neben den wissenschaftlichen Mitarbeitern nicht zuletzt auch die Professoren und Mitarbeiter aus Verwaltung und Technik beigetragen haben, werde ich für immer in guter Erinnerung behalten. Mein Dank gilt daher allen ehemaligen und aktuellen Mitarbeitern des IHE, mit denen ich während dieser Zeit zusammenarbeiten durfte.

Meinem Hauptreferenten, dem Institutsleiter des IHE, Prof. Thomas Zwick danke ich für die Möglichkeit am IHE zu promovieren. Wann immer ich seinen Rat gesucht habe, stand mir seine Tür offen. Neben vielen guten Ideen zum Fortgang meiner Arbeit waren es vor allem Energie und Motivation, die ich aus Gesprächen mit ihm mitgenommen habe. Bei allem, was ich während meiner Promotion am IHE gemacht habe, konnte ich mir seiner Unterstützung sicher sein, wofür ich dankbar bin.

Bei Prof. Ingmar Kallfass bedanke ich mich für die Übernahme des Korreferates, sowie die fortwährende fachliche Unterstützung während meiner Promotion. Auch seine Tür stand mir jederzeit offen – vielen Dank hierfür.

Für ihre Teilnahme an meiner Prüfungskommission möchte ich mich bei Prof. Ivers-Tiffée, Prof. Doessel und Prof. Puente-Leon bedanken.

Sebastian Diebold und Stefan Beer danke ich für ihre konstruktive und rasche Durchsicht meines Manuskripts.

Diese Promotion entstand im Rahmen eines Forschungsprojekts des IHE in Zusammenarbeit mit dem Fraunhofer-Institut für Angewandte Festkörperphysik (IAF) in Freiburg, finanziert von der Deutschen Forschungsgemeinschaft (DFG). Allen Mitarbeitern des IAF in Freiburg, ohne deren Unterstützung diese Promotion so nicht möglich gewesen wäre, gilt mein Dank. Namentlich erwähnen möchte ich an dieser Stelle insbesondere Arnulf Leuther, für seine Unterstützung in unserem gemeinsamen DFG Projekt, sowie Sandrine Wagner und Hermann Massler für die vielen Stunden, die sie beim Charakterisieren meiner Schaltungen verbracht haben.

Meinen ehemaligen Zimmerkollegen Andreas Lambrecht, Piotr Laskowski, Sebastian Diebold und Heiko Gulan gilt mein Dank für die gute Arbeitsatmosphäre in unserem Zimmer. Zahllose fachliche und nicht so fachliche Gespräche und Diskussionen werden mir für immer in guter Erinnerung bleiben. Mit ihnen als Zimmerkollegen war es leicht, jeden Tag zur Arbeit zu kommen.

Ingmar Kallfass und Thomas Zwick gilt mein Dank dafür, dass sie mir ermöglicht haben, sechs Monate meiner Promotion bei Agilent Technologies in den USA zu verbringen. Erst durch ihre aktive Unterstützung beim Herstellen des Kontakts, sowie bei der Organisation und Umsetzung, wurde dieser Auslandsaufenthalt möglich.

Nicht zuletzt und insbesondere gilt mein Dank meiner Familie. Sie ist der Grundpfeiler meines Lebens und ihr Anteil am Erfolg dieser Promotion ist daher mit nichts zu vergleichen. Vielen Dank, dass ihr immer für mich da seid.

Im März 2015,
Philipp Pahl

Zusammenfassung

Die vorliegende Arbeit beschreibt Konzepte für das Design von breitbandigen und effizienten, impedanztransformierenden Leistungsteilern auf Basis von verteilten, magnetischen Transformatoren, sowie deren Einsatz in Leistungsverstärkern bis in den oberen Millimeterwellenfrequenzbereich.

Um Richtfunkstrecken mit immer größeren Datenraten und Radarsysteme mit immer höherer Auflösungen zu erzielen, werden diese Systeme zu immer höheren Betriebsfrequenzen hin entwickelt. Durch die höhere Frequenz lassen sich mit den gleichen Designkonzepten, wie sie bei niedrigeren Frequenzen Anwendung finden, größere absolute Bandbreiten und somit letzten Endes höhere Datenraten bzw. Genauigkeiten erreichen. Ein Schwachpunkt dieser Systeme ist allerdings deren beschränkte Ausgangsleistung. Durch die Optimierung der zum Einsatz kommenden Transistoren hin zu immer größeren Operationsfrequenzen, sinkt deren Leistungsfähigkeit hinsichtlich der Ausgangsleistung. Durch die Parallelschaltung mehrerer Transistoren mittels geeigneter Designkonzepte lässt sich dieses Problem verringern. Die zum Einsatz kommenden Anpass- und Verteilnetzwerke reduzieren hierbei allerdings die Effizienz und/oder Bandbreite, die erreicht werden kann. Der Hauptgrund hierfür ist das mit steigender Anzahl paralleler Transistoren zunehmende Impedanztransformationsverhältnis. Der in der Literatur am Häufigsten verwendete Anpassnetzwerktyp sind LC-Netzwerke, die diskrete oder verteilte, passive Elemente (z.B. diskrete Kapazitäten, Induktivitäten, verteilte Transmissionsleitungen) verwenden um mehrere parallele Transistoren gleichzeitig an die Systemimpedanz anzupassen. Neben LC-Netzwer en verwenden siliziumbasierte Technologien (CMOS, SiGe) magnetische Transformatornetzwerke für die Leistungsaufteilung und Impedanztransformation. Der Hauptvorteil magnetischer Transformatornetzwerke gegenüber LC-Netzwerken ist die prinzipielle Unabhängigkeit ihrer Effizienz und Bandbreite vom notwendigen Impedanztransformationsverhältnis. Wie in der Literatur für siliziumbasierte integrierte

Schaltungen gezeigt, lassen sich bis in den Millimeterwellenfrequenzbereich hinein sehr effiziente Leistungsverstärker auf Basis von verteilten magnetischen Transformatoren realisieren.
Während das Konzept verteilter magnetischer Transformatoren bisher hauptsächlich auf CMOS und SiGe Prozessen vorteilhaft eingesetzt wurde, wird es in dieser Arbeit für das Design von sehr breitbandigen PAs auf einem III-V Halbleiterprozess eingesetzt. Zum Einsatz kommt dabei die 50 nm bzw. 35 nm metamorphe High Electron Mobility Transistor (mHEMT) Technologie des Fraunhofer Institut für Angewandte Festkörperphysik (IAF).
Die Einleitung erläutert die Motivation dieser Arbeit und gibt eine Einführung in die Funktionsweise von verteilten magnetischen Transformatoren sowie den Stand der Technik im Bezug auf deren Design. Kapitel 2 beschäftigt sich mit der Optimierung von magnetischen Transformatornetzwerken bezüglich ihrer Effizienz und Bandbreite. Es wird gezeigt, dass sich ein verteilter, magnetischer Transformator auf eine Verkettung von mehreren Transformatoreinheitszellen zurückführen lässt. Für die Einheitszellen lässt sich ein diskretes Ersatzschaltbild aufstellen. Basierend auf dem präsentierten Ersatzschaltbild werden Gleichungen aus der Literatur hergeleitet und evaluiert mit denen sich eine Einheitszelle eines verteilten, magnetischen Transformators hinsichtlich Effizienz optimieren lässt. Des Weiteren werden neue Designgleichungen entwickelt mit denen sich eine Optimierung hinsichtlich symmetrischer BW erreichen lässt. Da der verteilte, magnetische Transformator zwar kleiner aber nicht sehr klein gegen die geführte Wellenlänge ist, wird der Einfluss der Verteilung auf die mittels diskretem Ersatzschaltbild gewonnenen optimierten Transformatoreinheitszellen untersucht. Aufbauend auf diesen Untersuchungen wird das diskrete Ersatzschaltbild erweitert um die auftretenden Abweichungen in der Effizienz zwischen wirklichem, verteiltem Transformator und Ersatzschaltbild zu beseitigen. Basierend auf dem neuen Ersatzschaltbild werden neue Designgleichungen für die Effizienzoptimierung von verteilten, magnetischen Transformatoren entwickelt. Abschließend wird untersucht inwieweit die Optimierung einer Transformatoreinheitszelle sich auf einen kompletten verteilten Transformator, bestehend aus mehreren Einheitszellen, übertragen lässt. In Kapitel 3 wird eine Designstrategie für transformatorbasierte Anpass- und Verteilnetzwerke vorgestellt. Die bisher in der Literatur verwendeten Layouts von verteilten magnetischen Transformatoren werden diskutiert und ein

neues Layout vorgeschlagen, das vor allem auf dem verwendeten III-V Halbleiterprozess mit zwei Verdrahtungsebenen deutliche Vorteile bietet. In verteilten magnetischen Transformatoren werden anstelle von ineinander gewundenen Induktivitäten gekoppelte Leitungen verwendet. Basierend auf der Evaluation von verschiedenen Geometrien für gekoppelte Leitungen sowie unter Einsatz der Optimierungsgleichungen aus Kapitel 2 werden mehrere komplette Anpass- und Verteilnetzwerke auf Basis verteilter magnetischer Transformatoren in Kapitel 3 designt. Die erfolgreiche Optimierung bezüglich Effizienz und symmetrischer Bandbreite wird dabei mittels elektromagnetischer Simulationen der kompletten Netzwerke nachgewiesen. Anschließend werden in Kapitel 4 zwei Leistungsverstärker bei 98,5 GHz und 235 GHz demonstriert. Beide Verstärker verwenden das in Kapitel 3 vorgeschlagene, neue Transformatorlayout in ihren Eingangs- und Ausgangsnetzwerken und übertreffen bezüglich ihrer Bandbreite den aktuellen Stand der Technik deutlich. Im abschließenden Kapitel 5 werden die Schlussfolgerungen und neuen Erkenntnisse der Arbeit erläutert und diskutiert, sowie die neu entstandenen Optimierungs- und Designstrategien für das Design von verteilten magnetischen Transformatoren zusammengefasst.

Abstract

This dissertation describes concepts for the design of broadband and efficient, impedance transforming power dividers/combiners, based on distributed magnetic transformers. Besides a theoretical analysis of the transformers, their advantageous application in integrated millimeter-wave power amplifiers is presented.

In order to enable radio links with constantly increasing data rates and radar sensors with constantly higher resolutions, the operation frequency of those systems moves towards higher frequencies. By using higher operation frequencies, the same design concepts used at lower frequencies provide a wider absolute bandwidth and therefore higher data rates and better resolutions, respectively. One bottleneck of such systems, however, is their limited output power. Due to the optimization of the applied transistor technologies towards higher operation frequencies, the output power provided by a single transistor device decreases. To mitigate this issue, multiple transistors are used in parallel by means of a suitable impedance transforming power distribution design concept. On the other hand, the applied passive matching and distribution networks diminish the efficiency and bandwidth of the designed power amplifiers. The main reason for this is the required impedance transformation ratio, which increases with the number of parallel transistors. LC networks are the most common approach to implement such networks. They consist of lumped (i.e. inductors, capacitors) or distributed (i.e. transmission lines) passive elements, which are connected in a way that results in the desired impedance transformation and power distribution among the transistors. Besides LC networks, silicon based technologies (e.g. CMOS, SiGe) use magnetic transformers to act as power divider/combiner and impedance transformer simultaneously. The primary advantage of magnetic transformers compared to LC networks is that in a first approximation their efficiency and bandwidth are independent of the required impedance transformation ratio between the reference impedance of the power amplifier and its active elements i.e. transistors.

As demonstrated in literature, the concept of magnetic transformers can be applied to develop very efficient power amplifiers up to the millimeter-wave frequency regime. To date, the concept of distributed magnetic transformers has been mainly used on CMOS and SiGe processes. However, this thesis demonstrates that the concept can be beneficially applied to a III-V compound semiconductor process as well. All simulation and measurement results as well as analyses are based on Fraunhofer Institute of Applied Solid State Physic's (IAF's) 50 nm and 35 nm metamorphic High Electron Mobility Transistors technology.
The introduction explains the motivation for this work and presents the state of the art in terms of the design of distributed magnetic transformers. Furthermore, the general idea behind the concept is revised. Chapter 2 deals with the optimization of magnetic transformer networks regarding their efficiency and bandwidth. It is shown that a distributed magnetic transformer can be reduced to a cascade of multiple transformer unit cells. A lumped equivalent circuit model of these unit cells is presented and used to revise an optimization strategy and optimization equations used in literature to optimize magnetic transformers regarding their efficiency. Based on the same unit cell circuit model, novel optimization equations that result in the maximum symmetrical bandwidth are presented. Even though, the unit cell of a distributed magnetic transformer is smaller than the guided wave length, it is not much smaller. Hence, the influence of distribution on the frequency response of a unit cell optimized based on a lumped equivalent circuit model is evaluated. Based on the results of this evaluation, the existing lumped equivalent circuit model is extended, to eliminate the deviation between a real, distributed transformer and the lumped equivalent circuit model. Originating from the novel lumped model, new design equations for the optimization of distributed transformer unit cells regarding their efficiency are derived. At the end of chapter 2 the significance of the optimization of a single transformer unit cell for the behavior of an entire distributed transformer network consisting of multiple unit cells is investigated. Especially the amplitude and phase balance between the various ports of the transformer is evaluated. In chapter 3 a design strategy for magnetic transformer based impedance matching and power distribution networks is introduced. The standard layouts for such transformers are discussed and a novel layout, which proves to be more suitable for the applied two top-metal-layer III-V semiconductor process, is proposed. In distributed magnetic transformers,

coupled lines are used as magnetically coupled inductors. Based on the evaluation of various coupled transmission line geometries and the optimization equations derived in chapter 2, several complete impedance matching power distribution networks are designed in chapter 3. The successful optimization of the networks, with respect to their efficiency and bandwidth, is verified by electromagnetic field simulations. In chapter 4 two power amplifiers operating at 98.5 GHz and 235 GHz are demonstrated. Both amplifiers apply the novel transformer layout proposed in chapter 3 for their input and output matching networks and surpass the state of the art in terms of power amplifier bandwidth substantially. Finally, chapter 5 discusses and summarizes the novel design and optimization strategies proposed in this thesis.

Contents

List of Abbreviations and Symbols

Abbreviations

AC	Alternating Current
ADS	Agilent's Advanced Design System
airMS	air Microstrip
balun	balanced unbalanced transformer
BCB	Benzocyclobuten
BiCMOS	Bipolar Complementary Metal Oxide Semiconductor
BW	Bandwidth
CG	Common Gate
CS	Common Source
CMOS	Complementary Metal Oxide Semiconductor
CPW	Co-planar Waveguide
DAT	Distributed Active Transformer
DC	Direct Current
DT	Distributed Transformer
DUT	Device Under Test
ECM	Equivalent Circuit Model
EM	Electro Magnetic
FET	Field Effect Transistor
GaAs	Gallium Arsenide
GaN	Gallium Nitride
GND	Ground
HBT	Hetero Bipolar Transistor
HD	High Definition
HEMT	High Electron Mobility Transistor

IAF	Fraunhofer Institute for Applied Solid State Physics
IL	Insertion Loss
IMN	Input Matching Network
InP	Indium Phosphite
MAG	Maximum Available Gain
Met1	Metal 1
MetG	Galvanic Metal
mHEMT	metamorphic High Electron Mobility Transistor
MIM	Metal Insulator Metal
MMIC	Monolithic Millimeter-wave Integrated Circuit
mmW	Millimeter-Wave
MS	Microstrip
OMN	Output Matching Network
OS	Open-Short
PA	Power Amplifier
PAE	Power-Added Efficiency
RL	Return Loss
SEM	Scanning Electron Microscope
Si	Silicon
SiN	Silicon Nitrite
SiGe	Silicon Germanium
SNR	Signal-to-Noise Ratio
TEM	Transverse Electromagnetic
trafo	Transformer
TWA	Traveling Wave Amplifier
UAV	Unmanned Aerial Vehicle
vGND	Virtual Ground
www	world wide web

Constants

$c_0 = 299792458\,{}^{m}/_{s}$	speed of light
$e = 2.71828...$	Euler's number

j	$= \sqrt{-1}$	imaginary unit
π	$= 3.14159...$	ratio of a circle's circumference to its diameter

Greek Symbols

α	attenuation coefficient of a transmission line
β	wave number
$\varepsilon_{r_{e,o}}$	effective characteristic permitivity of the even, odd mode of a symmetrical coupled line
γ	propagation coefficient
Γ	reflection coefficient
η	efficiency
η_{max}	maximum achievable efficiency
λ	wavelength
λ_0	free space wavelength
ω	circular frequency ($= 2\pi f$)
ω_0	design/optimization circular frequency ($= 2\pi f_0$)
$\Delta\Phi$	phase balance
σ	conductivity
σ_{RCS}	radar cross section
τ	impedance transformation ratio

Latin Symbols

ΔA	amplitude balance
C	capacitance
C'	shunt capacitance per unit length
C_{eq}	equivalent series capacitance for the real parallel capacitance and load resistance
C_{ds}	drain source capacitance
C_p	capacitance at the primary side of the transformer
C_s	capacitance at the secondary side of the transformer

G	conductance
G'	shunt conductance per unit length
G_T	gain of the transmitter antenna
G_R	gain of the receiver antenna
I	current
I_p	current of the primary side of the transformer
I_s	current of the secondary side of the transformer
I_Σ	current sum
L	inductance
L'	Series inductance per unit length
L_p	self-inductance of the primary side of the transformer
L_s	self-inductance of the secondary side of the transformer
M	mutual inductance
N	number of elements
N_o	noise power
N_{order}	filter order
N_{or}	noise power at the receiver
P	signal power
$P_{1..N}$	port 1 ... port N
P_{1dB}	power at the 1 dB compression point
P_{diss}	dissipated power
P_{load}	power delivered to the load
P_{LR}	power loss ratio
P_{out}	output power
P_R	signal power at the receiver
$P_{R,min}$	minimum detectable receiver power
P_T	transmitted signal power from the transmitter
Q	quality factor
Q_m	quality factor/ ratio between the mutual inductance and mutual resistance
Q_p	quality factor of the transformer's primary side
Q_s	quality factor of the transformer's secondary side
R	resistance
R'	series resistance per unit length

R_0	optimization/design resistance
R_c	ratio between the voltages and currents of the c-mode on the two coupled lines of an asymmetrical coupled line
R_{eq}	equivalent series resistance for the real parallel capacitance and load resistance
R_L	load resistance
R_π	ratio between the voltages and currents of the π-mode on the two coupled lines of an asymmetrical coupled line
R_p	resistance of the primary side of the transformer
R_s	resistance of the secondary side of the transformer
U	voltage
U_{ds}	drain source voltage
U_{gs}	gate source voltage
U_Σ	voltage sum
$\hat{V}$	peak voltage
Z_0	characteristic impedance of a line
$Z_{0,d}$	characteristic impedance of an artificial drain line of a TWA
$Z_{0,g}$	characteristic impedance of an artificial gate line of a TWA
Z_c	characteristic impedance of the c-mode of an asymmetrical coupled line
Z_{k0}	scaling factor of the characteristic impedance of a symmetric coupled line
Z_o	odd characteristic impedance of a symmetrical coupled line
Z_e	even characteristic impedance of a symmetrical coupled line
Z_π	characteristic impedance of the π-mode of an asymmetrical coupled line
$\overline{\frac{Z_c+Z_\pi}{2}}$	mean value of the weighted characteristic impedance of both lines of an asymmetrical coupled line
c	propagation velocity
$cwidth$	geometrical width of the two lines of a coupled line
f_0	design/ optimization frequency
f_{max}	maximum oscillation frequency
f_t	maximum unity current gain cut-off frequency

gnd_sep	separation between the pair of coupled lines and ground to both of its sides
g_{m}	transconductance
k_{loss}	attenuation coefficient of the Butterworth filter
k_{m}	magnetic coupling
l	length of a coupled line
n	transformer turn ratio
opt	optimized
r	range (distance)
Δr	rcsolution

Math Symbols

$+$	addition
$-$	subtraction
$\frac{x}{y}$	x divided by y
x^*	complex conjugate value of x
$\sqrt{x}$	square-root of x
$\sqrt[n]{x}$	n^{th} root of x
$\sqrt{x}$	square-root of x
$\mathrm{Im}\{x\}$	imaginary part of x
$\mathrm{Re}\{x\}$	real part of x
X^{-1}	inverse of matrix X
$\arctan x$	arctangent of x
$\sinh x$	hyperbolic sine of x
$\tan x$	tangent of x
$\tanh x$	hyperbolic tangent of x
$\coth x$	hyperbolic cotangent of x
$\|x\|$	absolute value of x
$\frac{dU}{dx}$	U differentiated with respect to x
$\frac{\partial U}{\partial\{x,y\}}$	U partial differentiated with respect to x and y

1 Introduction

In today's world, information is one of the most important goods and people are used to having it available everywhere and all the time. To access information usually laptops or smart phones with various types of wireless interfaces are used. While the majority of accessed services today (e.g. email, www, etc.) only demand a rather small data rate, it is the growing number of participants and services that leads to a steadily increasing demand for higher data rates. This trend will speed up in the future as new services, that require higher data rates, like wireless HD video streaming, high speed file transfer or real time active services e.g. HD video telephony and wireless gaming become more common [YXVG11, Wir10]. Shannon's equation [Sha49] reveals that the capacity of a wireless channel[1] depends on the available bandwidth (BW) and signal-to-noise ratio (SNR):

$$C = BW \cdot \log_2 \{1 + SNR\} = BW \cdot \log_2 \left\{1 + \frac{P}{N_o}\right\} \tag{1.1}$$

For a communication link the ratio between the signal power (P) and noise power (N_o) at the receiver is relevant. Using the Friis transmission equation [Fri46, GW98] the SNR at the receiver (SNR_R) can be expressed as a function of the transmitter output power (P_T):

$$SNR_R = \frac{P_R}{N_{o,R}} = \frac{P_T G_T G_R \left(\frac{\lambda_0}{4\pi r}\right)^2}{N_{o,R}} \tag{1.2}$$

By observing Eqn. 1.1 and 1.2, it becomes obvious that the demand for higher data rates and therefore a higher link capacity demands the transmission of broadband high power signals.

[1] The channel is only limited by additive white Gaussian noise.

In the scenarios mentioned above, at least one of the two parties involved in the wireless transmission is usually a mobile device, which puts constraints on the form factor of the transceiver components used to establish the wireless connection. For instance, the RF front-end in all smart phones available today is implemented as one or multiple monolithic microwave integrated circuits that accommodate the multitude of required RF components within the smallest possible space. Based on very small and compact Monolithic Millimeter-wave Integrated Circuits (MMICs), data transmission with a high absolute bandwidth in the high millimeter-wave (mmW) frequency regime has been demonstrated experimentally [KAS+11, HMM11]. However, the relative bandwidth of these systems is rather small. One of the reasons for this is the lack of available output power over a wide bandwidth [AKLD+13, LDKT+12, ABM+14].

Besides wireless communication, radar and imaging applications also benefit from bandwidth and output power. In Eqn. 1.3 the relation between the resolution and the bandwidth (BW) of a radar is given [Sko03]:

$$\Delta r = \frac{c}{2BW} \tag{1.3}$$

If the bandwidth increases, the minimal separation (Δr) necessary to distinguish two objects in range decreases as well, which leads to a higher resolution.
The maximum range at which a radar system can still detect a target depends on the output power (P_S) of the radar transmitter. By rearranging the radar equation, the maximum range of a radar is defined as [Sko03]:

$$r_\mathrm{max} = \sqrt[4]{\frac{P_\mathrm{T} G_\mathrm{T} G_\mathrm{R} \lambda^2 \sigma_{RCS}}{(4\pi)^3 P_\mathrm{R,min}}} \tag{1.4}$$

where G_T and G_R represent the gain of the transmitter and receiver antenna, respectively, σ_{RCS} is the radar cross section of the target and $P_\mathrm{R,min}$ stands for the minimum detectable receiver power.
As for communication transceivers integrated in mobile devices, the demand for very compact and small radar transceivers results in a growing research interest in highly integrated MMIC based radar transceivers [SUC12, PJA12, BPJ+13]. Such sensors can be implemented in a lot of different devices where size and

weight matters, e.g. power tools, smart phones, unmanned aerial vehicles (UAV), etc. By increasing the operation frequency towards high mmW frequencies, the absolute bandwidth and therefore the resolution is increased. However, as for communication transceivers published in literature, the output power available over a wide frequency range is still limited to date.
In conclusion it can be noted that by moving the carrier frequency of communication and radar front-ends towards higher frequencies, high data rates and high resolution radar images can already be achieved. However, one bottleneck of these systems to date is the lack of output power over a wide bandwidth, which limits the range and therefore the usability of these systems.
Besides radar and communication systems, RF measurement equipment benefits from the availability of high output power over a wide bandwidth, too. High power levels usually result in a higher dynamic range [TWK98] and if the systems can cover a wider frequency range, their field of application increases as well.

1.1 Technology limitations

Even though passive diode based technologies can be found in literature, published MMICs operating at high millimeter-wave frequencies mostly apply active transistor technologies. Compared to diode based technologies, MMICs based on fast transistors are usually superior in terms of output power on the transmitter and noise figure on the receiver side of a mmW transceiver chip.
Fig. 1.1 shows the current state of the art in power amplifier (PA) MMICs with respect to their output power. PAs are usually applied as the last component of a transmitter, right before the antenna and therefore indicate the highest output power available on the various MMIC technologies.
By comparing the different technologies (e.g. Complementary Metal Oxide Semiconductor (CMOS), Silicon Germanium (SiGe) Hetero Bipolar Transistor (HBT), Indium Phosphite (InP) HBT, InP High Electron Mobility Transistor (HEMT), Gallium Arsenide (GaAs) metamorphic High Electron Mobility Transistor (mHEMT), Gallium Nitride (GaN) HEMT), III-V compound semiconductors turn out to be superior compared to Silicon (Si) based technologies regarding their output power. All technologies however demonstrate the same trend of output power versus

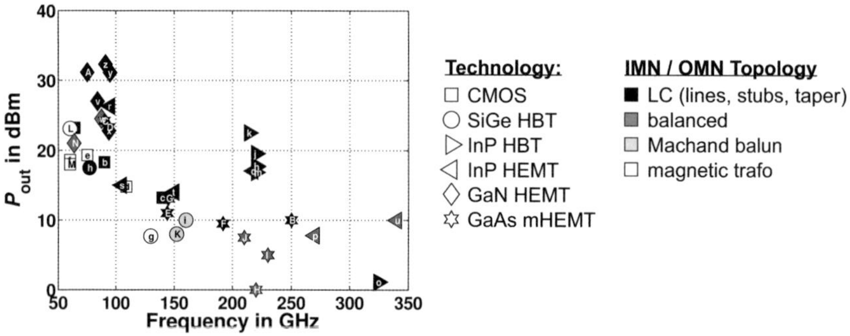

Figure 1.1: State of the art in output power for millimeter-wave power amplifier design.

frequency: $P_{\text{out,dB}} \sim {}^{1}/_{f}$.

In terms of output power, one limiting factor in all high speed MMIC technologies is the low on-state breakdown voltage (V_{bds}) of the available transistors [Lee04]. To provide enough gain at high mmW frequencies, the gate length of field effect transistors (FETs) or the emitter width of bipolar transistors is scaled down. Furthermore, the compound semiconductors are modified in their composition towards higher electron mobility and maximum drift velocity, which leads to a reduction of their band gap. Both measures increase the cut-off frequency (f_{t}) but reduce the breakdown voltage of the technology [LW00].

Today's III-V [STK^{+}11] and Si based [RHF12] semiconductors achieve maximum oscillation frequencies (f_{max}) of >900 GHz and >500 GHz with breakdown voltages of about 1.5 V. Thus, the voltage swing at the drain or collector of the transistor may not exceed ±0.75 V (neglecting the knee voltage). Without an impedance transformation between the active device and the common 50 Ω output impedance, this results in a maximum power transfer of only 5.6 mW into the load.

$$P_{\text{load}} = \frac{\hat{V}^2}{2 \cdot R_{\text{L}}} \tag{1.5}$$

To allow for more output power from a single Si based or III/V transistor, an impedance transformation is necessary. In general high-speed transistor devices provide low voltage but high current swings and therefore a low output impedance [ALK+08, AKRH02b]. By using an impedance transformation network the 50 Ω load can be transformed into a smaller load impedance at the drain or collector of the transistor, which according to Eqn. 1.5 results in a higher output power. Since the maximum drain and collector currents cannot be increased indefinitely (e.g. the maximum drain current of an advanced 35 nm mHEMT with a total gate width of 2 x 30 µm in common-source (CS) configuration is 1600 mA/mm [STK+11]), there is usually an optimum load resistance which leads to the highest possible output power, associated with the highest possible voltage and current swing product [Cri06]. For instance at 200 GHz, simulations with Fraunhofer Institute of Applied Solid State Physics's (IAF's) design kit for their 100 nm technology reveal an optimal load impedance of approximately 15 Ω for a 4 x 15 µm mHEMT in CS configuration. This means that the impedance transformation network has to transform the 50 Ω load into 15 Ω at the transistor, which corresponds to an impedance transformation ratio ($\tau = \frac{R_{\mathrm{PA,out}}}{R_{\mathrm{trans.,out}}}$) of more than three. Implemented as simple second order LC network, the matching network is fairly narrow banded (Fig. 1.2). If the order of the matching network is increased, the bandwidth can be increased but at the expense of higher losses. In standard LC matching networks this issue becomes even worse, if multiple active elements are used in parallel, as it is usually done to overcome the maximal power limitation of a single transistor device by means of power combining. Since putting multiple transistors in parallel means putting their optimal drain/collector load impedance in parallel, the transformation ratio between the required load at the parallel transistors and the common 50 Ω load at the PA output port gets even higher. As with the example of the four finger mHEMT with a total gate width of 60 µm (4 x 15 µm mHEMT), it results in a transformation ratio of more than 12, if four transistors are put in parallel.
In terms of output power, the main problem of a high transformation ratio is the low quality factor (Q) of the monolithic integrated passive components (e.g. transmission lines, lumped inductors/capacitors, etc.), which do not achieve values as high as those of PAs operating at lower frequencies, whose matching networks can be implemented in hybrid technologies for instance. The issue of the high

necessary transformation ratio in combination with the relatively high losses of the passive components in MMICs also aggravates the design of broadband power amplifiers, since their matching networks are usually more complex than those optimized for optimal impedance transformation only over a small frequency range. Hence, broadband LC matching networks suffer even more from the high losses of the passive elements.

In conclusion, it can be said that the design of power amplifiers operating at high millimeter-wave frequencies requires the use of advanced transistor technologies, which are optimized for very high cut-off frequencies. A single transistor device in such a process, even operated at its optimum load, only provides a very limited amount of output power. Therefore, it becomes necessary to use multiple transistors in parallel to achieve enough power. The parallelization of multiple devices however leads to a required impedance transformation ratio, even higher than the already high ratio, necessary to match a single device to the output impedance of the PA. As the ratio increases so does the loss of standard LC networks, which are mainly used at high millimeter-wave frequencies (Fig. 1.3). This issue gets even worse, if the PA is supposed to be broadband, since more complex and therefore more lossy LC networks are necessary to provide the optimal impedance over a wide bandwidth.

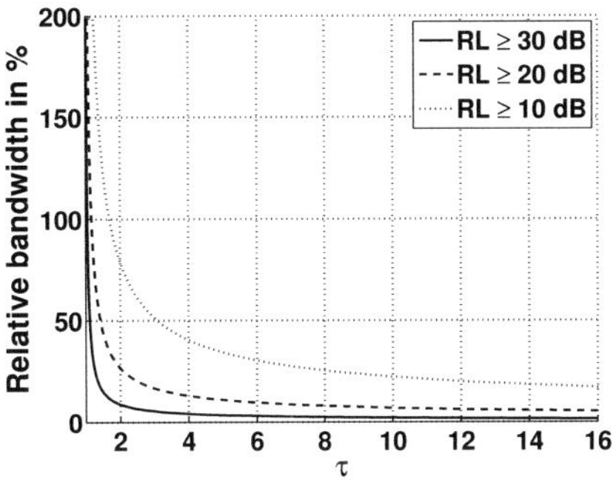

Figure 1.2: Simulated symmetrical bandwidth of an ideal second order LC network for various return loss conditions versus the impedance transformation ratio.

1.2 State of the art in mmW power divider combiner structures for PAs

Fig. 1.3 shows the current state of the art in MMIC PA design between 50 GHz and 350 GHz. The maximum output power (P_{out}), power-added efficiency (PAE) and relative -3 dB bandwidth are plotted in three different figures. To associate the points, each referenced amplifier has a unique letter, which can be found in all three plots. If the PAE or the -3 dB bandwidth is neither mentioned in the referenced paper nor can be determined from any other given information, the point is omitted in the respective plot. The different technologies are illustrated by different marker shapes and the color coding identifies the type of applied power divider/combiner matching network. Appendix A gives a list of references associated to the state of the art plot.

The most common concept for the design of impedance transformers and power combiner/divider networks in PAs operating at high mmW frequencies is the standard LC matching. Based on distributed lines of various characteristic impedances in combination with T-junctions, the power division and combination is achieved.

PAs using GaN HEMT technologies together with LC networks achieve the highest reported output powers (P_{out}) [BBC$^+$11, NMM$^+$10] below 100 GHz. For instance at 91 GHz a P_{out} of 32.3 dBm has been reported [BBC$^+$11]. The relative bandwidth of most of these PAs however is rather small. In general it is between 10 % and 20 %.
In the mmW frequency range, the highest reported relative BW is achieved by an InP HEMT PA [SL01]. Based on LC networks the PA shows a relative BW of 57 % at a center frequency of 105 GHz and an output power of 15 dBm.
Very recently, also, PAs based on CMOS technology have been reported that use LC nctworks of high order (multiple lines of various impedances - low Q network) to achieve a wide bandwidth [HTL$^+$13]. At 64 GHz, 90 GHz and 140 GHz relative bandwidths of 39 %, 29 % and 21 % are demonstrated. The output power of these PAs is among the state of the art for CMOS PAs.
Above 110 GHz the majority of published PAs is based either on HBTs or HEMTs, which all apply hetero-junctions to achieve high cut-off frequencies [SL12]. PAs using InP HBTs in combination with LC networks (tapered lines) have demonstrated output powers of more than 20 dBm at around 220 GHz [RGR$^+$13, RRG$^+$12a, RRG$^+$11], which are the highest reported power values for single MMICs operating beyond 100 GHz to date. The highest relative bandwidth among these PAs however is no more than 22 %.
More broadband MMIC power amplifiers have been demonstrated with quadrature hybrid couplers (eg. branch-line, coupled line, Lange coupler, etc.) as power dividers.
A GaN HEMT based PA operating at 87.5 GHz, which uses Lange couplers to divide the power between two parallel branches of four consecutive CS stages is presented in [MKM$^+$12]. It achieves an output power of 24.5 dBm over a relative bandwidth of 34 %.
For balanced MMIC PAs above 100 GHz, the highest reported bandwidth is demonstrated by an InP HEMT PA operating at 270 GHz. The PA applies branch-line couplers as power divider and combiner and shows a relative bandwidth of 23 % at an output power of 7.8 dBm [DMR$^+$08].
Another PA power divider/combiner concept, which can be found in literature, uses Marchand balanced unbalanced transformers (baluns). The differential signals provided by this approach are particularly interesting for Si based processes (e.g.

CMOS, SiGe), where substrate losses can be mitigated by the use of virtual grounds (vGNDs). SiGe HBT based PAs with an output power of 10 dBm and a relative bandwidth of 6 % at 160 GHz have been demonstrated with this concept [SCP13]. This approach also facilitates the design of broadband PAs, as shown, for instance, by a SiGe PA operating at 152 GHz, which exhibits a relative bandwidth of 23 % and an output power of 8 dBm [SHP13].
Instead of Marchand baluns, magnetic transformer baluns are also used as simultaneous balun and power divider. Especially CMOS and SiGe PAs below 100 GHz apply the concept of Distributed Active Transformers (DATs) to enable highly efficient PAs, whose efficiency is much higher than that based on standard LC networks. For instance, [WCW12] presents a DAT based CMOS PA operating at 75 GHz with a PAE of 19.2 %, an output power of 19.3 dBm and a relative BW of 13 %. The PAE of this PA is among the state of the art not only for CMOS but MMIC PAs in general.
Unlike Marchand baluns, the DAT concept allows for more than two parallel branches and performs an impedance transformation inherently, which is why it is very efficient.
The transformer based PA operating at the highest frequency to date uses SiGe HBTs to achieve a P_{out} of 7.7 dBm over a relative bandwidth of 15 % at 130 GHz. Even though it appears that the relative bandwidth of transformer based PAs is less than that of PAs based on LC or hybrid coupler designs, this is not because of any BW limitation of the transformer concept in general. The BW in the published transformer based MMIC PAs is mainly limited due to their optimization for high output powers and efficiency (PAE) rather than broadband performance. The DAT concept, for instance, applies a minimum number of passive matching elements, which limits the chances of achieving a flat gain shape over a wide bandwidth. The low cut-off frequencies of Si based transistor technologies, which result in less device gain compared to III-V HBTs and HEMTs are also disadvantageous for a broadband design. A single passive transformer alone enables relative bandwidths of 100 % or more, as will be demonstrated in this thesis. The published PAs by now, however, do not harness this property, because of the reasons mentioned above.

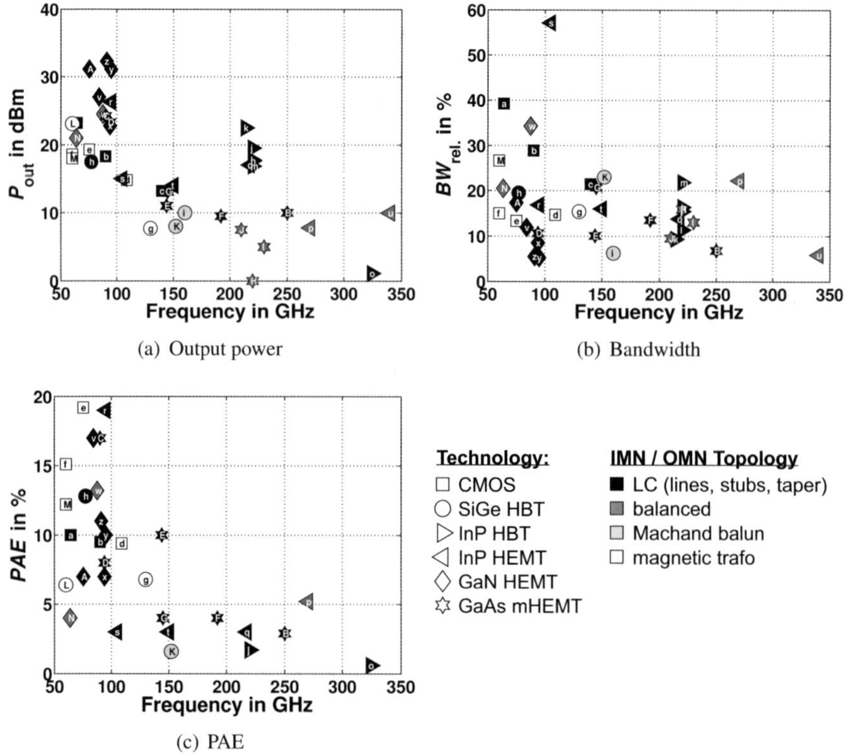

(a) Output power

(b) Bandwidth

(c) PAE

Figure 1.3: State of the art in millimeter-wave power amplifier design.

1.3 State of the art in millimeter-wave transformer design

This thesis will introduce a concept for the design of highly efficient, broadband, transformer based Input Matching Networks (IMNs) and Output Matching Networks (OMNs) operating at mmW frequencies and benefiting from the high relative BW and efficiency that can be achieved with the Distributed Transformer (DT) concept. To familiarize the reader with this concept and to explain its method of operation, this section will give a brief introduction alongside an overview of the state of the art in DTs used for matching networks of mmW PAs.

Section 1.2 presented the current state of the art in PA design. It revealed that most of the amplifiers based on III-V semiconductor processes apply standard LC transformation networks to achieve the output power, PAE and BW that marks the current state of the art at mmW frequencies. On the other hand PAs built on Si based processes (CMOS, SiGe Bipolar Complementary Metal Oxide Semiconductor (BiCMOS)) tend to focus not only on LC but also magnetic transformer based impedance transformation and power divider/combiner networks. These networks offer an efficient method to achieve high output power on Si based technologies.

The most prominent approach of a magnetic transformer, to be used at mmW frequencies, is the DAT, which was first proposed in 2001 by Aoki et al. [AKRH01]. Its major advantage compared to standard transformer based impedance matching networks is its ability to combine the output signal of multiple transistors while performing an impedance transformation at the same time. In [AKRH02b] the principle operation of a DAT is introduced. The concept mainly relies on differential push-pull transistor pairs and inductive coupling. By connecting multiple differential push-pull transistor pairs by means of slab inductors, in a way that their drain/collector currents sum up, an artificial primary inductor loop is built if they are arranged in a circle. Since all transistor pairs connected via slab inductors are operated with 180° phase shift, a uniform circular alternating current (AC) at the fundamental frequency is generated, which excites an alternating magnetic field. The alternating magnetic field induces a voltage into a secondary inductor loop, which is positioned inside the artificial primary magnetic loop, i.e. the loop built by the slab inductors. While the current on each primary inductor slab and

the secondary inductor loop are identical but with opposite phases, the voltage is summed up on the secondary loop (four 1:1 transformers in series on the secondary and in parallel on the primary side - Fig. 1.4(b)). This leads to an impedance transformation ratio of 1:4 between each primary and the secondary loop and a power combination at the same time.
Figure 1.4(a) depicts the general geometry as it is proposed in [AKRH02b].

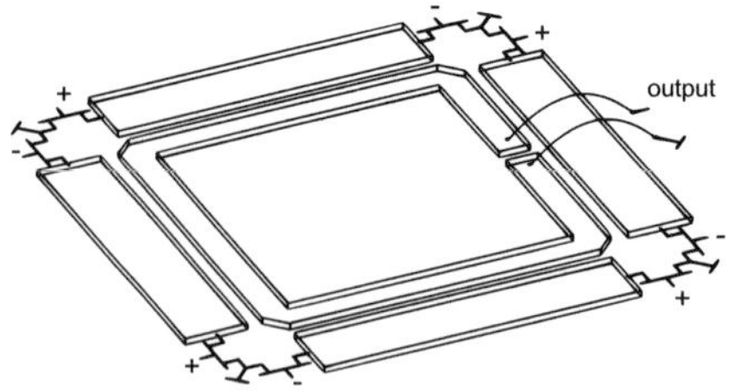

(a) Principle geometrical design of the DAT proposed by Aoki et al. in [AKRH02b].

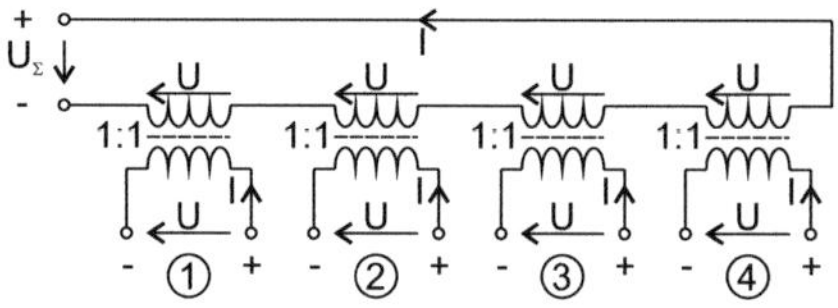

(b) Principle schematic of the DAT.

Figure 1.4: Principle geometrical design and schematic of the DAT proposed by Aoki et al.

The first PA with more than 1 W of output power at 2.4 GHz on a 0.35 µm - CMOS process was demonstrated in 2001 [AKRH01] by means of the DAT concept.
A theoretical comparison between impedance transforming, power divider/combiner networks based on LC resonant, standard magnetic transformer and DAT structures is given in [AKRH02a]. The main conclusion of this analysis is that the efficiency of an output power combining matching network based on the DAT concept, unlike LC networks, is in principle independent of the impedance transformation ratio. This is an important advantage, due to the low output impedance of transistor devices operated at mmW frequencies (section 1.1).
Compared to standard magnetic transformer based impedance transformers, which are also independent of the transformation ratio, the DAT has its advantage in its ability to perform an impedance transformation and power combining at the same time. Due to the lower loss of slab inductors compared to spiral inductors, used in standard impedance transformers, the DAT approach is also more efficient.

Moreover, typical standard transformers cannot be easily applied for an impedance transformation network at mmW frequencies because of an unrealistic high turn ratio (N), which would be required for high power generation, i.e. power matching ($N = \sqrt{\tau}$). The low Q of multi-turn spiral inductors and the low magnetic coupling prevent their efficient usage [ALK$^+$08, AKRH02a]. Therefore, the *distribution* of a high turn ratio by means of multiple transformers with a reduced turn ratio, as it is done in the DAT concept, is more practical for the integrated design.
The first DATs that were used for efficient micro- and millimeter-wave PA design on Si based substrates, were all arranged in a way, that their primary slab inductors build an artificial primary loop [AKRH01, AKRH02b, JR05, PKKH07, PHKH08]. One problem that comes with this configuration however is that the IMN, which has to provide the differential input signals to all push-pull transistor pairs, has to be routed either through the middle of the DAT (shortest way) or around it (long way). Both of these approaches come with specific disadvantages. The short way through the middle of the DAT has the disadvantage, that the alternating magnetic field of the DAT induces a signal into the IMN, which means that there is some sort of unwanted positive or negative feedback, that can cause instability [AKRH02b, PMEvR12]. To mitigate this effect grounded *guard lines* in parallel to the lines of the IMN can be used [KLL$^+$05]. The advantage of the short way is the short line length, leading to less loss than for the long way solution. On the other hand, the long way solution is less influenced by unwanted feedback, since the alternating magnetic field is less intense outside the DAT. The downside of this solution however is a quite complex and space consuming routing to deliver the necessary differential signals to the gates/bases of the transistors. For instance, the V-band amplifier published in [PG07], which has an IMN that goes around the DAT in the middle, has an IMN that occupies 400 times the space of its DAT based OMN. Besides the large amount of wafer space, the long transmission lines also result in high losses of about 5 dB for the IMN of this PA. An alternative geometry for a DAT is proposed in [HLN06] and shown in Fig. 1.5. The so called *figure 8 transformer layout* possesses single inductive loops on the primary and secondary side of the transformer. As can be seen in Figure 1.5, this geometry allows for a more efficient connection of the push-pull transistor pairs, as it avoids the need to route the IMN either through the DAT or around it. Based on this or quite similar geometries, multiple power amplifiers have been

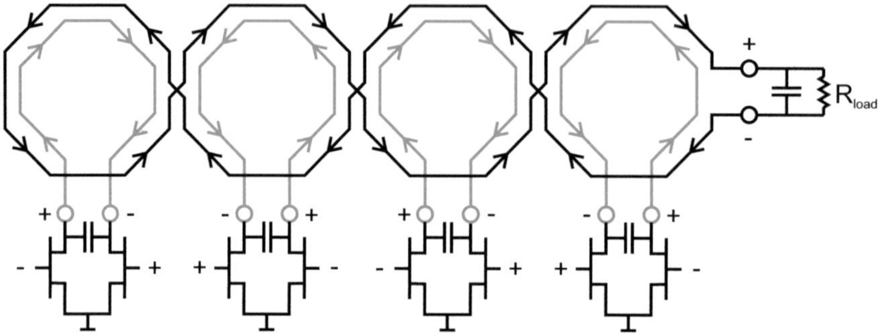

Figure 1.5: Principle geometry of a DAT applying a *figure 8* geometry proposed by Haldi et al. in [HLN06].

reported in literature [LLN06, HCR+08, EMPvR11]. In general it can be said, that all proposed geometries for the design of DATs to date are composed of one or multiple loops on either the primary, the secondary or both sides of the transformer.
Depending on the available process metal layer stack, mainly three different kinds of coupling configurations between the primary and secondary side of the DAT can be found in literature. Their principle geometries are depicted in Fig. 1.6.

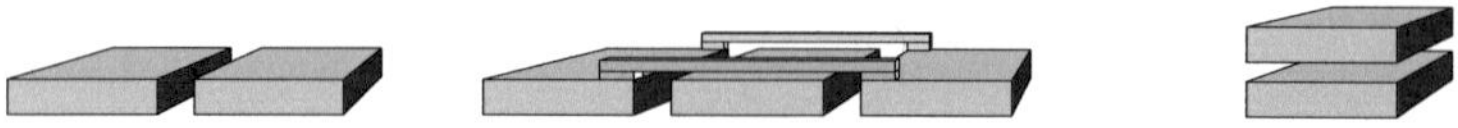

(a) edge coupled (b) double edge coupled (c) broadside coupled

Figure 1.6: The most common coupling schemes in DATs.

As already described in [AKRH02a], the magnetic coupling between the primary and secondary side of the DAT is essential to achieve an efficient transformer. While the first published DATs used standard edge coupled coupling schemes (Fig. 1.6(a)), the need to improve the coupling was addressed quite soon after the introduction of the DAT. In [KLL+05] an optimized DAT geometry is presented, which, in addition to the extraneous primary loop slabs of a standard edge coupled transformer, uses additional primary slabs at the inside of the secondary inductor loop (double edge coupled - Fig. 1.6(b)). This coupling scheme results in a higher coupling coefficient and therefore a higher efficiency of the transformer. If more than one metal layer is available a stacked inductor configuration (Fig. 1.6(c)) can lead to an even higher coupling factor as has been shown, for example, in [PG07, JTHW09].

The first DATs, which all applied a circular or *figure 8* geometry, were mainly designed by extensive Electro Magnetic (EM) simulations. Since the primary and secondary side of the transformer were designed to be either an artificial (multiple primary slabs) or real inductor loop (secondary side), this was necessary to account for the coupling between opposite sides of the inductor. To overcome the necessity for a full wave EM simulation of the DAT, an inductor configuration with two broadside coupled loops above a ground shield with side bars (grounded parallel lines) was proposed in [PGFR05, PG07] (Fig. 1.7(a)). The ground shield limits the expansion of the alternating magnetic field and therefore reduces the mutual coupling between opposite sides of the inductors. Applying the ground shield makes every side of each loop independent of all other sides of the loop. Therefore, it becomes possible to build the entire DAT out of multiple sections of coupled line unit cells (Fig. 1.7(b)). By reducing the DAT to a number of connected coupled lines, a scalable model of the DAT can be set up. The DATs designed in [PGFR05, PG07] were developed by means of such a model. However, a detailed description of the applied models is not given.

The reason why the DAT has the term *active* in its name is because usually part of the transistors output capacitance is used to compensate the inductance of the transformer. If necessary, an additional parallel capacitor is added to provide the desired load impedance at the transistor output, but most of the compensation is usually done by the transistor's drain-source/collector-emitter capacitance. Therefore, one part of the DAT is actually the transistor with its parasitic output elements.

Unlike DATs, DTs, as defined in this thesis, do not incorporate the output capacitance of an attached transistor into the transformer, but compensate the complete parasitic inductance by means of additional parallel capacitors. There are a couple of published DTs at micro- and millimeter-wave frequencies [PKKH07, LTLW07,

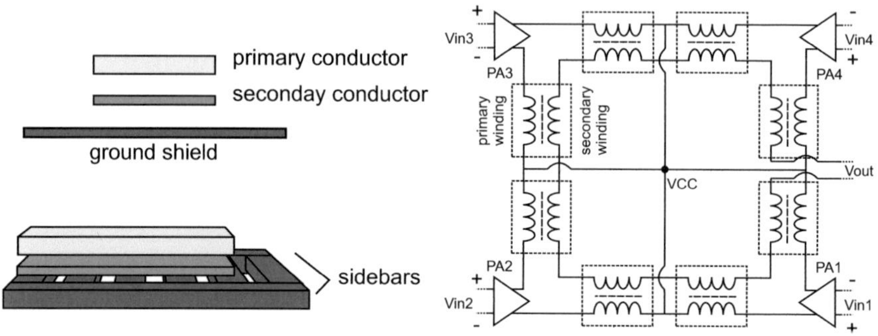

(a) Coupled line unit cell model.

(b) Schematic of a DAT composed out of multiple independent unit cell elements.

Figure 1.7: DAT design based on independent unit cell elements as proposed in [PGFR05, PG07].

JTHW09, TTB+13] which all provide differential *purely resistive* low impedance ports on the primary side and usually a single-ended 50 Ω port on the secondary side. In these designs, the matching of the transistor devices is usually achieved by additional matching networks of various complexity.

Besides transformer based OMNs, there are also publications that apply DATs or DTs as part of their IMNs, where the concept can be applied as well [LTLW07, CRN09, HLR10, CN11, GXC12, TTB+13, OKH13]. In this case the 50 Ω input impedance of the PA's input port is transformed into multiple low impedance ports, whose impedance is close to optimum power matching for the connected transistor devices. The general principle is exactly the same as for DAT/DT applied in an OMN.

Besides a series connection of multiple 1:1 transformers it is also possible to connect the secondary side of a DAT in a parallel manner as shown, for instance, in [LVG10, ALD+12, TTB+13, OKH13]. In this case, the power is not combined by adding the voltage on the secondary side of the transformer (*voltage combiner* - Fig. 1.8(a)), but by adding the currents on the different parallel branches (*current combiner* - Fig. 1.8(b)). This, however, leads to an inverse impedance transformation compared to the standard voltage combiner. A comparison between voltage combiners and current combiners with respect to their application as DATs/DTs in PA circuits can be found in [GXC12].

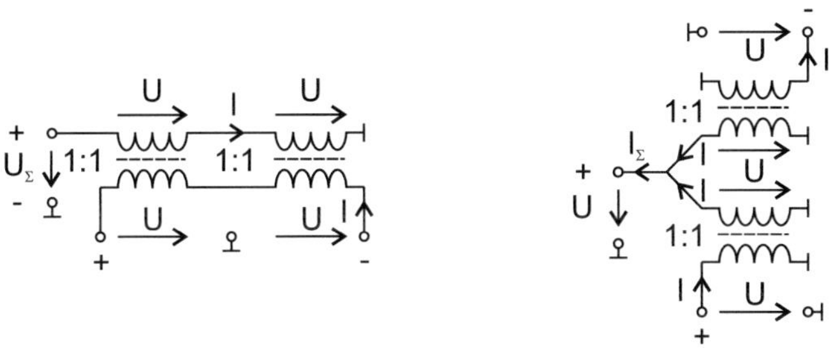

(a) voltage combiner

(b) current combiner

Figure 1.8: DT/DAT in voltage and current combiner configuration.

To conclude this brief introduction of the fundamental principle and the state of the art in DAT and DT design, an overview of power amplifiers based on the DAT/DT concept and with respect to output power (P_{out}), PAE and relative BW is depicted in Fig. 1.9. As for the state of the art in mmW PA design, appendix B contains a list of references associated to the state of the art plot shown in Fig. 1.9. It is noteworthy to say that to date all published DAT and DT based power amplifiers are built on Si based processes (e.g.: CMOS, SiGe). Due to the fact that the Q of integrated passive components (inductors, capacitors, and transmission lines) used for impedance matching on CMOS and SiGe processes is worse than on its III-V counterparts [Lon00], the investigation of new methods for efficient power combining and impedance matching has been propelled on Si. Using III-V semiconductor processes, the loss of the applied IMNs and OMNs is still acceptable at low millimeter-wave frequencies, however, the exact same need for alternative impedance matching and power combining concepts also arises for III-V semiconductors as the frequency increases.

It has already been stated that the focus of almost all published mmW PAs on

Si based technologies is on output power and efficiency, which is why their relative bandwidth is rather small. However, the DT concept in general allows for a very wide bandwidth, if the transistor technology provides enough gain to design broadband gain cells at the operation frequency. This, for instance, is demonstrated by two CMOS PAs operating at 3 GHz and 48 GHz with a bandwidth of 4 GHz ($BW_{\text{rel.}} = 133\,\%$) [HTLW10] and 22 GHz ($BW_{\text{rel.}} = 45\,\%$) [CRN08], respectively.

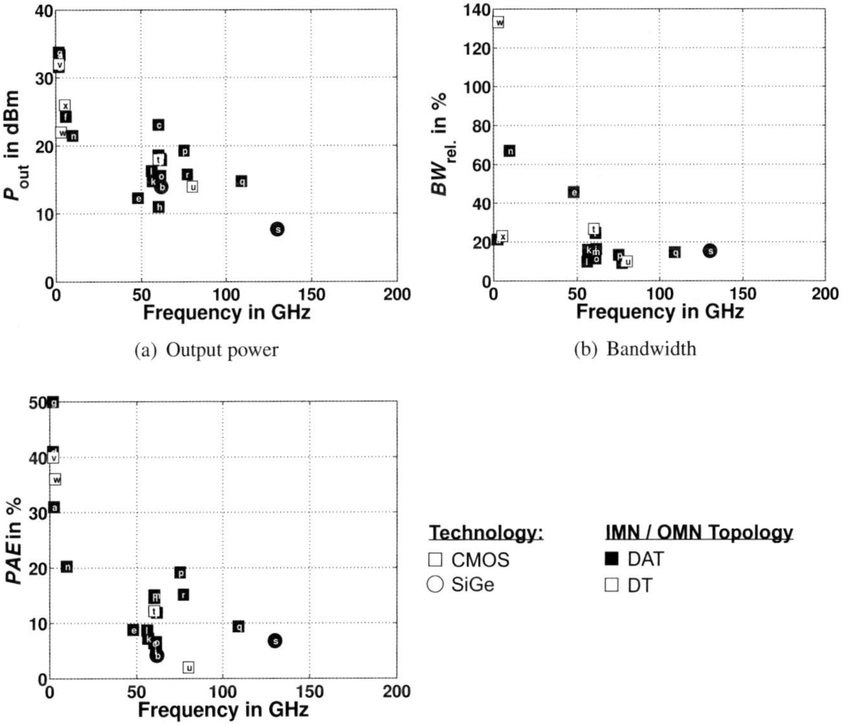

(a) Output power

(b) Bandwidth

(c) PAE

Figure 1.9: State of the art in magnetic transformer based millimeter-wave power amplifier design.

1.4 Problem statement and outline of the thesis

One reason why MMIC based communication and radar systems at mmW frequencies have not had their breakthrough yet, is the limited available output power of such systems. The state of the art plot shown in section 1.2 reveals that besides output power, in particular the available bandwidth of the reported PAs is still small.

Based on LC networks, III-V compound semiconductor technologics achieve the highest output power and bandwidth over the entire mmW frequency regime. Si based technologies like CMOS and SiGe, however, also allow for high output power and bandwidth up to about 160 GHz. Besides LC networks, Si based PAs apply matching networks based on magnetic transformers to overcome the limitations imposed by the technology (e.g. limited output power and gain of the active devices and high loss of the passive components). The main advantages of magnetic transformer networks over LC networks are that their efficiency and bandwidth in principle are independent of the impedance transformation ratio of the network. Furthermore, magnetic transformers are very compact and can be used as balun, without any major modifications.

Instead of multi-turn transformers, PAs on Si usually use DT/DATs, since the required impedance transformation ratio of PA matching networks would lead to multi-turn transformers with unrealistic high turn ratio and low quality factor. The DT/DAT concept on the other hand facilitates the design of very efficient matching networks.

Up to now mmW DAT/DT networks are optimized for maximum efficiency, i.e. maximum output power. In particular, the DAT concept, which uses the parasitic capacitance of the transistors to compensate for the unavoidable inductance of the magnetic transformer network, concentrates on efficiency by using as little passive matching elements as possible.

Since all published DT/DAT based PAs use CMOS or SiGe technologies, the availability of virtual grounds between parallel transistors is heavily used to minimize losses. This results in the fact that almost all published PAs use some sort of circular DT/DAT geometry, where neighboring transistors can be combined to push-pull transistor pairs.

The prevailing strategy for the design of DT/DATs to date is based on extensive EM simulations. Usually, the geometry of one slab inductor is chosen based on EM simulations of the entire slab geometry. The strategy is to maximize the coupling coefficient and quality factor of the transformer to maximize the efficiency of the network. The sizes of the inductors and parallel capacitors of the transformer are either chosen to result in a convenient size of the DT/DAT or their values are determined from the efficiency optimization of a simple transformer model. This simple equivalent circuit model (ECM) will be the starting point of the theoretical analysis conducted in this thesis and therefore will be described in more detail in chapter 2.

Even though, DT/DAT matching networks are only used for Si based mmW PAs by now, there is no reason why the concept should not be applicable for the design of highly efficient and very broadband mmW PAs on III-V compound semiconductor technologies. Since the optimal input and output impedance of transistors decreases with an increase of operation frequency, the required transformation ratio increases as well. The high transformation ratio results in limited efficiency and bandwidth of standard LC networks, not only on Si but also III-V technologies. Both, efficiency and bandwidth of a magnetic transformer based network, however, are independent of the transformation ratio, which makes it an interesting candidate on all technologies.

III-V processes achieve higher cut-off frequencies and output power than Si-based technologies. Therefore, the combination of DT based matching networks with advanced III-V compound semiconductor technology, should lead to PAs, which extend the current state of the art in terms of output power and efficiency over a wide relative bandwidth.

To develop a design scheme, which enables such broadband PA designs is the objective of this thesis and it translates into the following primary sub-tasks, that will be engaged in the chapters ahead:

- Theoretical analysis of a DT not only in terms of efficiency but also bandwidth.

- Development of a design procedure and layout for DT based mmW matching networks on a III-V semiconductor process.

- Verification of the theoretical findings based on the novel layout.
- Application of the novel DT layout for the design of PAs that prove the superiority of the approach by a comparison with the state of the art in mmW PA design.

The outline of each chapter is as follows:
In **chapter 2** the theoretical basis for the optimization of a DT network will be given. The DT will be reduced to a cascadation of multiple DT unit cells, which will be optimized in terms of efficiency and bandwidth. In literature, DT/DAT design equations for maximum efficiency, based on a lumped ECM, can be found. This model and the related design equations will be introduced and new design equations, that facilitate the design of DTs with a maximum symmetrical bandwidth, will be developed.
Furthermore, it will be demonstrated that the standard ECM used in literature for the efficiency optimization of DT/DAT networks to date, omits the severe influence of resistive coupling between the primary and secondary side of the distributed magnetic transformer. Thus, if the simplistic optimization equations from literature are applied, the optimization of the DT results in incorrect parameter/geometry values. To overcome this problem, a new ECM and new design equations for the optimization of DT/DATs for maximum efficiency are presented. Since the proposed optimization of DT/DATs is based on the optimization of unit cells, the deviation between the frequency response of a single DT unit cell and the entire DT, consisting of multiple optimized unit cells, is evaluated at the end of the chapter.
Chapter 3 will demonstrate a design strategy for DT networks that uses the optimization equations for maximum efficiency or maximum symmetrical bandwidth developed in chapter 2. Based on the design rules of Fraunhofer IAF's 35 nm and 50 nm mHEMT technology, a novel DT layout for two top-metal-layer III-V semiconductor processes will be proposed. Starting with the evaluation of various possible coupled line configurations, whose EM simulations are the basis for the optimization of the DT unit cells, multiple complete DT based IMN are designed and evaluated by EM simulations. The good agreement between the optimization and the EM simulation of the entire IMN as well as the results themselves, will not only demonstrate the applicability of the design strategy, but also the excellent

performance that can be achieved with the DT concept at high mmW frequencies on III-V processes. To conclude chapter 3, simulations and measurements of a 1:4 port DT matching network will be presented.
In **chapter 4** the DT layout introduced in chapter 3 will be applied for the design of two mmW PAs. Due to the very broadband impedance transformation of the DT based IMNs and OMNs in combination with two different broadband gain cells in between, both presented amplifiers clearly exceed the state of the art in terms of relative bandwidth at their operation frequency. Relative -3 dB bandwidths of 98 % at a center frequency of 98.5 GHz with an output power of about 15 dBm as well as a relative 3 dB bandwidth of 29 % for a PA operating at 235 GHz with an output power of approximately 5 dBm are shown.
The findings of this work will finally be summarized and discussed in **chapter 5**.

2 Theoretical Analysis of a Distributed Transformer

The objective of this chapter is the development of measures that allow the design of DTs, either providing the lowest possible insertion loss (IL), i.e. highest possible efficiency or the widest possible symmetrical bandwidth for a given DT unit cell geometry at a given frequency. The principle geometry of a DT/DAT will be broken down to an ECM for a single transformer unit cell, which will be the starting point of all the theoretical investigations ahead.

The most common design strategy for achieving the lowest possible insertion loss is introduced and its applicability for the design of mmW transformer based IMNs and OMNs will be investigated. Since the analysis is based on a lumped ECM, the deviation between this model and the distributed implementation is investigated by replacing the lumped inductor/transformer by its coupled line counterpart. It will be shown that as soon as distributed loss is introduced into the analysis the standard procedure starts to fail, leading to a false optimization, i.e. wrong inductance and capacitance values and in the end an inefficient operation of the transformer.

Besides the optimization for the best efficiency at a given frequency, an approach for the design of DTs and DATs, yielding maximum symmetrical bandwidth at a specified frequency, will be presented. Unlike demonstrated in literature by most of the published Si based PAs at mmW frequencies, this approach can lead to very wide relative BWs of more than 100 %.

2.1 The DT as a composition of multiple single 1:1 transformers

In chapter 1 the state of the art of micro- and millimeter-wave power amplifiers based on the principle of DT/DAT was presented. As stated, most of the published transformer designs are developed by means of extensive EM simulations [CRN08, EMPvR11, KBH11, TTB+13]. To achieve an optimal solution some of the designs apply a basic ECM, which can be utilized to calculate the required inductor and capacitor values for the most efficient transformation for a given frequency [AKRH01, HCR+08, JTHW09]. The principle behind the DAT/DT concept was already described in chapter 1. In Fig. 2.1(a) the most generic approximation of a DT/DAT is shown.

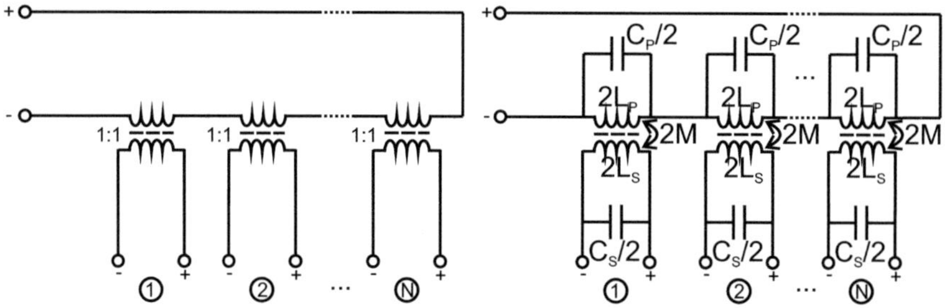

(a) Schematic of an ideal DT (N ideal 1:1 transformers).

(b) Schematic of a DT constructed out of N compensated 1:1 transformers.

Figure 2.1: Lumped DT model.

There, the DT is approximated by a number of ideal 1:1 transformers. These transformers are connected in series on the primary and in parallel on the secondary side, which in this case leads to the desired power division and impedance transformation from a higher impedance (Z_{IN}) on the primary side (i.e. 50 Ω) to a lower impedance of Z_{IN}/N on the secondary side. Here, N represents the number of ideal 1:1 transformers.

Since there are no ideal magnetic 1:1 transformers in reality, a more realistic schematic of the DT is the one depicted in Fig. 2.1(b). All ideal 1:1 transformers are replaced by mutually coupled inductors and capacitors in parallel. The capacitors are required since every real magnetic transformer has a parasitic inductance, that is unavoidable and has to be compensated to let the real 1:1 transformer look like an ideal one for a given frequency. Even though the DT is usually not composed out of lumped elements, the extension of the entire DT is supposed to be small compared to the wavelength, which to the first extent, makes it possible to describe it by lumped elements.

If the DT is used within an IMN, it usually transforms a high input impedance into multiple lower impedances at the output ports. The input impedance divides equally over the attached 1:1 transformers, which means that in principle it is sufficient to investigate and optimize a single 1:1 transformer to achieve an optimal solution. Each of these transformers is terminated with a real impedance of $\frac{Z_{IN}}{N}$ on both sides of the coupled inductors. Since each of the 1:1 transformers can be either seen as a two port device with two differential ports or a four port structure with 4 single-ended ports the schematic in Fig. 2.1(b) can be extended in a way that enables the extraction of a single-ended ECM for a DT unit cell. In Fig. 2.2 the relation between the unit cell model, which will be applied throughout this chapter and the schematic of a DT can be seen.

Based on the ECM, shown in Fig. 2.2(b), the equations for the optimization for maximum BW and maximum efficiency will be calculated. To determine the most efficient element values for a given frequency and load termination, the ECM will be extended by additional elements that represent the loss.

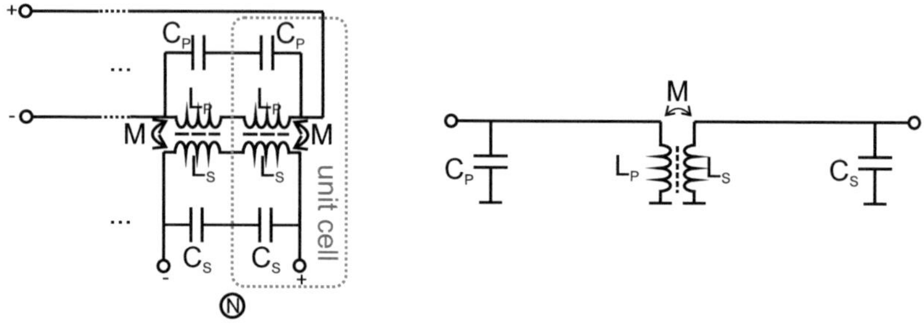

(a) Schematic of a DT (N 1:1 trafos) indicating the origin of the transformer unit cell.

(b) Schematic of a magnetic 1:1 transformer/DT unit cell.

Figure 2.2: From an ECM of a DT towards an ECM of a DT unit cell.

2.2 Efficiency optimization based on the lumped unit cell model

The most common approach to model the loss of a transformer at a given frequency is by adding resistors ($R_\mathrm{p}, R_\mathrm{s}$) in series to the inductors of the transformer ($L_\mathrm{p}, L_\mathrm{s}$). By doing this, the lossless schematic (Fig. 2.2(b)) is extended as shown in Fig. 2.3(a). In this ECM the inductances on the primary and secondary side of the 1:1 transformer are assumed to be identical, which is a valid assumption in almost all DAT/DT circuits. Therefore, the variable L will be used for the inductance of the primary inductor (L_p) as well as the secondary inductor (L_s), from here on.

To facilitate the calculation of optimum values for the inductors and capacitors of a transformer unit cell, the 1:1 transformer in Fig. 2.3(a) is replaced by its equivalent T-model as shown in Fig. 2.3(b) [Nik07].

The ECM of Fig. 2.3(b) is the origin of most of the optimization equations published in literature for DATs [HCR+08, AKRH02a]. The objective of the optimization is the determination of the ideal inductance L and capacitance C_s to minimize the loss and maximize the efficiency of the transformer for a given load termination and frequency. Therefore, the schematic used for the optimization

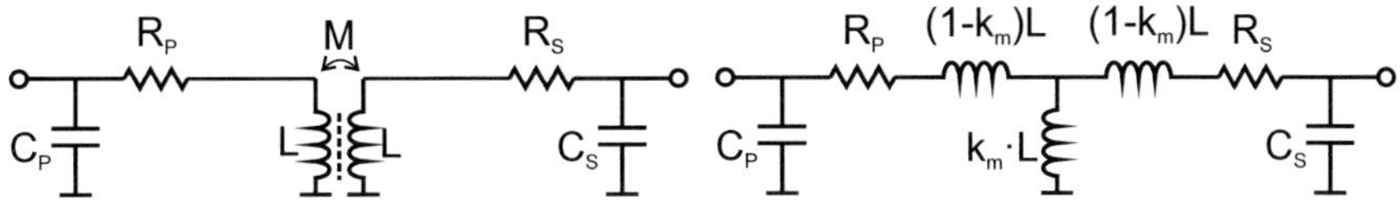

(a) Schematic of a lossy magnetic 1:1 transformer.

(b) Lumped element model of a lossy DT unit cell.

Figure 2.3: Lossy lumped element model of a single transformer unit cell.

has to include the load termination as well and is depicted in Fig. 2.4(a). Note that by applying this ECM, the efficiency of a DT unit cell is independent of the capacitor on the primary side (C_p), which is only used for matching purposes. In the ideal case, where the inductance and resistance of the primary and secondary side are identical, C_p equals C_s for an 1:1 transformer. If there is any inequality, both values will deviate as C_p will be usually used to resonate the net inductance on the primary side or to achieve the desired complex port impedance (power matching).
For the derivation of the efficiency of one transformer unit cell, it is convenient to replace the parallel capacitor and the resistive load by an equivalent series load and capacitance, as it is shown in Fig. 2.4(b).

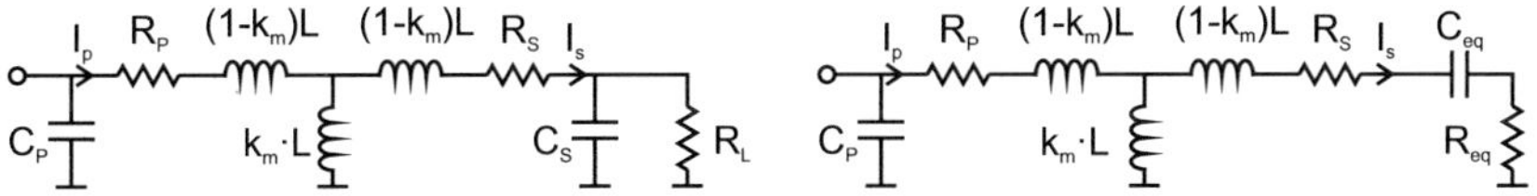

(a) Schematic of a lossy DT unit cell with load termination R_L.

(b) Schematic of a lossy DT unit cell with an equivalent series capacitance C_{eq} and load termination R_{eq}.

Figure 2.4: Lumped element model used for the efficiency calculation.

The relation between the frequency dependent R_{eq} and C_{eq} and the original parallel elements (C_{s}, R_{L}) is given by Eqn. 2.1.

$$\begin{aligned} R_{\text{eq}}(\omega) &= \frac{R_{\text{L}}}{1+(\omega R_{\text{L}} C_{\text{s}})^2} \\ C_{\text{eq}}(\omega) &= \frac{1+(\omega R_{\text{L}} C_{\text{s}})^2}{\omega^2 R_{\text{L}}^2 C_{\text{s}}} \end{aligned} \tag{2.1}$$

Based on the schematic in Fig. 2.4(b) the efficiency of a single DT unit cell can be expressed as shown in Eqn. 2.2

$$\eta = \frac{P_{\text{load}}}{P_{\text{load}}+P_{\text{diss}}} = \frac{R_{\text{eq}}}{R_{\text{eq}}+R_{\text{s}}+\left|\frac{Z_{\text{s}}+j\omega k_{\text{m}}L}{j\omega k_{\text{m}}L}\right|^2 R_{\text{p}}} \tag{2.2}$$

where $Z_{\text{s}} = j\omega(1-k_{\text{m}})L+R_s+\frac{1}{j\omega C_{\text{eq}}}+R_{\text{eq}}$ and R_{eq}, C_{eq} are frequency dependent (Eqn. 2.1).

By resonating the inductors $j\omega(1-k_{\text{m}})L$ and $j\omega k_{\text{m}}L$ on the secondary side using C_{eq} (Eqn. 2.3), the efficiency (η) can be maximized by minimizing the current (I_{p}) through resistor R_{p} [AKRH02a].

$$\omega L = \frac{1}{\omega C_{\text{eq}}} \tag{2.3}$$

To put the loss, which is represented by R_{p} and R_{s} in relation with the unit cell's primary and secondary inductance, quality factors are introduced:

$$Q_{\text{p}} = \frac{\omega L}{R_{\text{p}}} \qquad Q_{\text{s}} = \frac{\omega L}{R_{\text{s}}} \tag{2.4}$$

This leads to the following equation for the efficiency:

$$\eta = \frac{R_{\text{eq}}}{R_{\text{eq}}+\frac{\omega L}{Q_{\text{s}}}+\left|\frac{\frac{\omega L}{Q_{\text{s}}}+R_{\text{eq}}}{j\omega k_{\text{m}}L}\right|^2 \frac{\omega L}{Q_{\text{p}}}} \tag{2.5}$$

If Eqn. 2.5 is differentiated with respect to L, the optimum value for L, resulting in the highest efficiency at the design frequency (ω_0), can be determined to be:

$$\omega_0 L = \frac{\frac{1}{\sqrt{\frac{1}{Q_\mathrm{s}^2}+\frac{Q_\mathrm{p}}{Q_\mathrm{s}}k_\mathrm{m}^2}}}{1+\frac{1}{\frac{1}{Q_\mathrm{s}^2}+\frac{Q_\mathrm{p}}{Q_\mathrm{s}}k_\mathrm{m}^2}} R_\mathrm{L} \tag{2.6}$$

The corresponding C_s is:

$$C_\mathrm{s} = \frac{1}{2}\frac{R_\mathrm{L}+\sqrt{R_\mathrm{L}^2-4(\omega_0 L)^2}}{\omega_0^2 L R_\mathrm{L}} \tag{2.7}$$

which follows directly from Eqn. 2.1 and Eqn. 2.3.

Fig. 2.5 demonstrates the behavior of a unit cell, which is modeled as shown in Fig. 2.3(b). In these plots, both sides of the ECM are terminated with the same resistive load and the quality factors of the primary and secondary side of the transformer are assumed to be identical ($Q = Q_\mathrm{p} = Q_\mathrm{s}$). The magnetic coupling coefficient in each figure is $k_\mathrm{m} = 0.5$ and the optimization frequency is 240 GHz. Furthermore, C_p is determined to resonate the net inductance on the primary side, as will be the case in all efficiency optimizations ahead.

By investigating the various plots, it can be seen, that the optimization leads to the most efficient solution at 240 GHz for all quality factors. Depending on the loss, the efficiency of the network degrades, which is depicted in the efficiency (Fig. 2.5(c)) as well as the transmission coefficient plot (Fig. 2.5(d)).

While the design achieves an optimum result in terms of efficiency, it obviously does not achieve the highest possible symmetrical bandwidth as well. The first resonance of the ECM coincides with the optimization frequency, which can be seen in Fig. 2.5(b) and Fig. 2.5(d). A wider symmetrical bandwidth could be achieved by using different values for the elements of the ECM. How these values can be calculated, will be described in the next section of this chapter.

To evaluate the minimum loss involved with a transformer based IMN or OMN, the maximum efficiency in terms of the quality factors of the inductors as well as the coupling coefficient can be calculated by combining Eqn. 2.1, 2.6 and 2.7 together with Eqn. 2.5. The resulting equation for the maximum efficiency is:

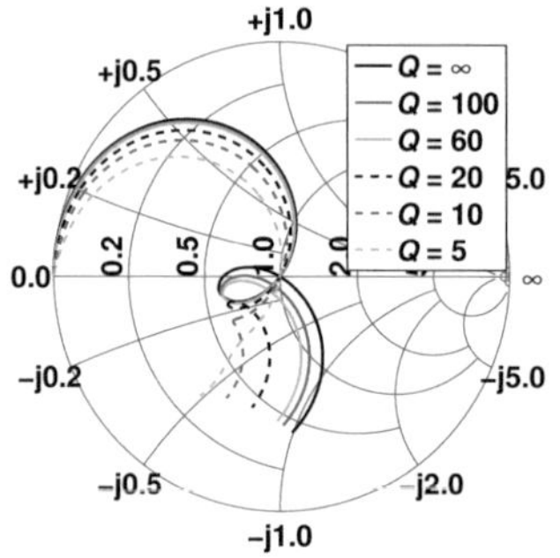

(a) Reflection coefficient

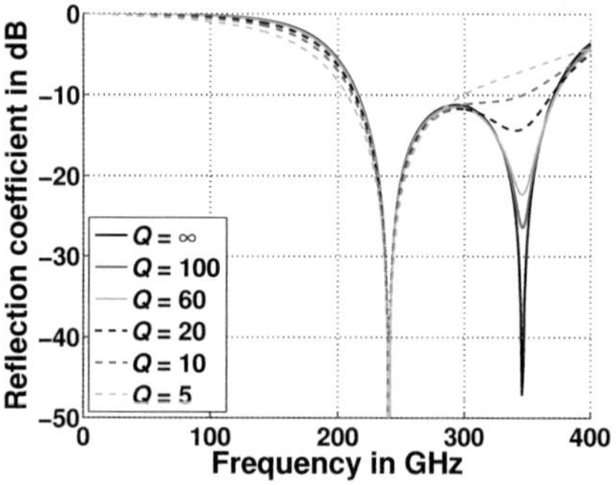

(b) Reflection coefficient

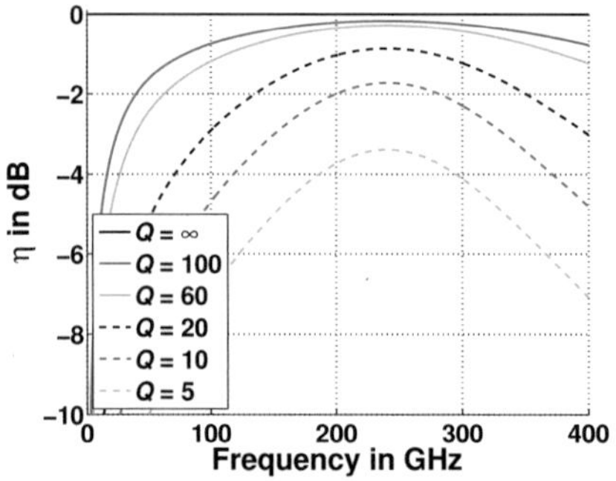

(c) Efficiency

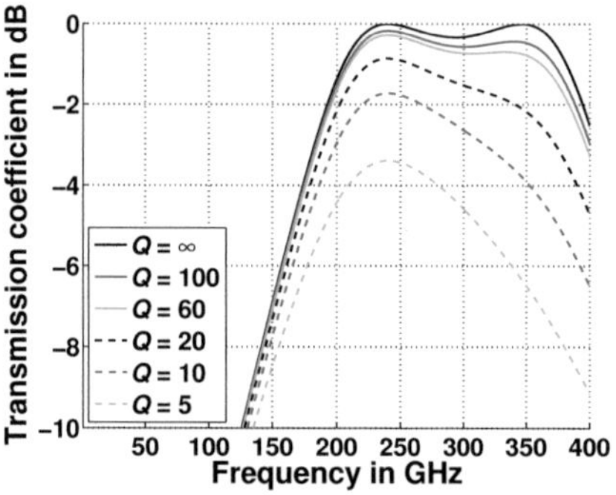

(d) Transmission coefficient

Figure 2.5: Influence of loss ($Q_\mathrm{p} = Q_\mathrm{s} = Q$) on the performance of an 1:1 transformer synthesized for maximal efficiency at a center frequency of 240 GHz and with $k_\mathrm{m} = 0.5$. The curves are obtained by evaluating the ECM depicted in Fig. 2.3(b), which is terminated with the same load impedance on both sides.

$$\eta_{max} = \frac{1}{1 + \frac{2}{Q_p Q_s k_m^2} + 2\sqrt{\frac{1}{Q_p Q_s k_m^2}\left(1 + \frac{1}{Q_p Q_s k_m^2}\right)}} \tag{2.8}$$

A contour plot of η_{max} versus the magnetic coupling and the quality factor is shown in Fig. 2.6. For simplicity and to facilitate visualization the quality factor of both the primary and secondary side's inductors are again assumed to be identical ($Q = Q_p = Q_s$).

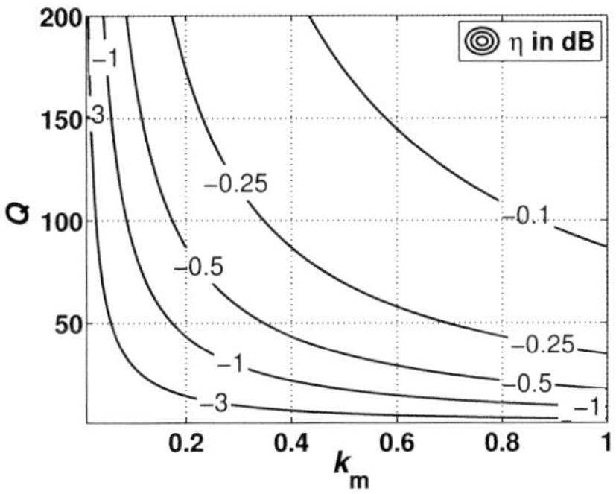

Figure 2.6: Maximum efficiency (η_{max}) in dB versus quality factor (Q) and magnetic coupling coefficient (k_m).

Fig. 2.6 illustrates that the overall efficiency of a magnetic transformer is limited by the magnetic coupling coefficient as well as the quality factors of the transformer. Therefore, to increase the efficiency, the objective for the design of a DT unit cell is to achieve the highest possible quality factors and the highest possible magnetic coupling between the primary and secondary side of the transformer.

2.3 Bandwidth optimization based on the lumped unit cell model

While the previous section derived the design equations for the highest possible efficiency at a given frequency, this section will provide design equations, that lead to the highest possible symmetrical bandwidth at a given frequency. Equations for the design of transformer unit cells that provide either a maximally flat, i.e. Butterworth IL or a return loss (RL) of better than 10 dB, 20 dB and 30 dB, will be presented. As for the derivation of the equations for the highest possible efficiency, the starting point for the investigation will be the ECM shown in Fig. 2.2(b).

While the internal transformer loss has to be modeled to enable an optimization in terms of efficiency, these losses play only a minor role, if one wants to design a DT unit cell with respect to its bandwidth around a given frequency. If resistors are included into the ECM, their influence on the BW of a DT unit cell is usually negligible, while the complexity of the design equations increases substantially. Therefore, the loss of the magnetic transformer will be neglected for the determination of the optimum transformer unit cell inductance and capacitance in terms of BW. To justify this approach, the influence of loss on the transmission and reflection coefficients of a DT unit cell, designed by the presented equations, will be verified afterwards.

The ECM of a DT as it is depicted in Fig. 2.2(b) can be regarded as a second order filter consisting of two coupled resonators. Since the schematic is completely symmetrical with respect to both ports, the inductor as well as the capacitor values of the resonators on the primary and on the secondary side have to be identical.

Therefore, the ECM, which will be used for the optimization has only two independent values, namely C and L. All other values are predetermined by either the geometry of the transformer, which determines the coupling coefficient (k_{m}), or the number of parallel ports, i.e. twice the number of 1:1 transformers in series, which determines the load termination R_{L}.

To derive the correct values of C and L in terms of k_{m} and R_{L}, the insertion loss method [Poz04] will be applied. This method is well known from filter theory and it defines the response of a filter in terms of its IL, which for a lossless filter is also called power loss ratio (P_{LR}):

$$P_{LR} = \frac{\text{Power available from source}}{\text{Power delivered to load}} = \frac{1}{1-|\Gamma(\omega)|^2} = \text{IL} \tag{2.9}$$

Here, the $\Gamma(\omega)$ represents the reflection coefficient. As the transformer is assumed to be lossless, the IL is only influenced by the RL, which is equal to $\frac{1}{|\Gamma(\omega)|^2}$.
Fig. 2.7 depicts the ECM of a single lossless DT unit cell. The model is terminated with the same load on both sides and is derived from the ECM shown in Fig. 2.2(b) by replacing the 1:1 transformer by its equivalent T-model.

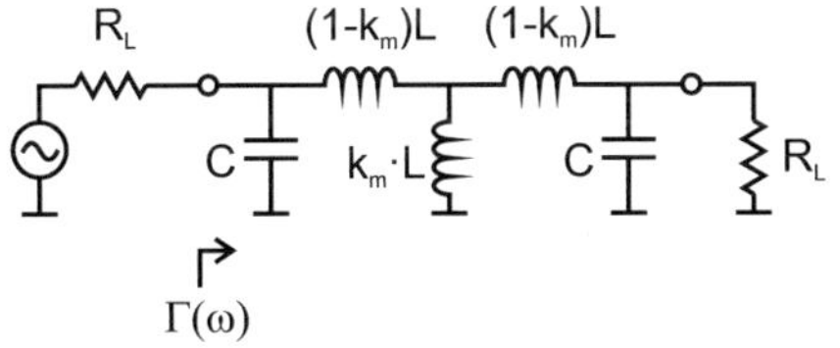

Figure 2.7: Lossless ECM of the DT unit cell applied for the calculation of P_{LR}.

If the reflection coefficient ($\Gamma(\omega)$) marked in Fig. 2.7 is calculated and inserted in Eqn. 2.9, P_{LR} can be derived to be

$$P_{LR} = \frac{1}{4}\frac{((1+R_L^2\omega^2C^2)\omega^2(1+k_m)^2L^2 - 2CR_L^2\omega^2(1+k_m)L + R_L^2)}{L^2k_m^2R_L^2\omega^2} \cdot \frac{((1+R_L^2\omega^2C^2)\omega^2(k_m-1)^2L^2 + 2CR_L^2\omega^2(k_m-1)L_p + R_L^2)}{L^2k^2R_L^2\omega^2} \tag{2.10}$$

for a single DT unit cell.
Based on P_{LR} an optimization of the unit cell with respect to various common filter responses (e.g. maximally flat/Butterworth, equal ripple/Chebyshev or linear phase/Bessel) is possible. In the case of a DT unit cell, the objective is to achieve the maximum bandwidth in terms of IL and RL for a given source and load termination (R_L) as well as a given magnetic coupling coefficient (k_m).
Hence, even though all kinds of standard filter response types can be applied for

the optimization, the most interesting ones are maximum flat and equal ripple responses, which will be discussed next.

2.3.1 Maximally flat (Butterworth) frequency response

The P_{LR} of a physically realizable maximally flat response bandpass filter has to have the following form [Poz04]:

$$P_{\mathrm{LR}} = 1 + k_{\mathrm{loss}}^2 \left(\frac{\omega - \omega_0}{\omega_{\mathrm{c}}} \right)^{2N_{\mathrm{order}}} \tag{2.11}$$

Here, N_{order} represents the order of the filter, whose passband extends from $\omega_0 - \omega_{\mathrm{c}}$ to $\omega_0 + \omega_{\mathrm{c}}$ around a center frequency of ω_0. At the edge of the passband ($\omega_0 - \omega_{\mathrm{c}}$, $\omega_0 + \omega_{\mathrm{c}}$) the P_{LR} becomes $1 + k_{\mathrm{loss}}^2$ and for $\omega = \omega_0$ the first $2N_{\mathrm{order}} - 1$ derivatives of the maximally flat P_{LR} are zero, which is the distinguished characteristic of a maximally flat response.

As described before, the two magnetically coupled resonators of a DT unit cell (Fig. 2.3(a)) equal a second order band pass filter. This property of the unit cell can also be seen in the derived P_{LR} (Eqn. 2.10), whose highest order of ω is four after expansion.

For the unit cell of a DT to behave like a second order maximally flat filter, the P_{LR} shown in Eqn. 2.10 has to be transformed in a way that it equals the P_{LR} of a second order Butterworth filter, which means that the first three derivatives of the P_{LR} have to be zero (Eqn. 2.12) at the center frequency, i.e. design frequency.

$$\left.\frac{\mathrm{d}P_{\mathrm{LR}}}{\mathrm{d}\omega}\right|_{\omega=\omega_0} \stackrel{!}{=} 0; \quad \left.\frac{\mathrm{d}^2P_{\mathrm{LR}}}{\mathrm{d}^2\omega}\right|_{\omega=\omega_0} \stackrel{!}{=} 0; \quad \left.\frac{\mathrm{d}^3P_{\mathrm{LR}}}{\mathrm{d}^3\omega}\right|_{\omega=\omega_0} \stackrel{!}{=} 0 \tag{2.12}$$

This condition can be applied to solve Eqn. 2.10 for the two unknown variables, which results in the following equations for L and C versus k_{m} and R_{L}:

$$L = -\frac{\sqrt{-2+2k_{\mathrm{m}}^2+2\sqrt{-k_{\mathrm{m}}^2+1}}\cdot R_{\mathrm{L}}}{(k_{\mathrm{m}}+1)(k_{\mathrm{m}}-1)\cdot\omega_0} \quad (2.13)$$

$$C = \frac{1-\sqrt{-k_{\mathrm{m}}^2+2-2\sqrt{-k_{\mathrm{m}}^2+1}}}{\sqrt{-2+2k_{\mathrm{m}}^2+2\sqrt{-k_{\mathrm{m}}^2+1}}\cdot\omega_0\cdot R_{\mathrm{L}}} \quad (2.14)$$

With Eqn. 2.13 and 2.14 it is now possible to design DTs or DATs that behave like a second order Butterworth filter in terms of their IL and RL.
To demonstrate the theoretical bandwidth, which can be achieved, the reflection coefficient versus frequency for various coupling coefficients is plotted in Fig. 2.8(a). Depending on the desired minimum return loss the relative bandwidth for a RL of better than 10 dB, 20 dB and 30 dB versus k_{m} is depicted in Fig. 2.8(b).

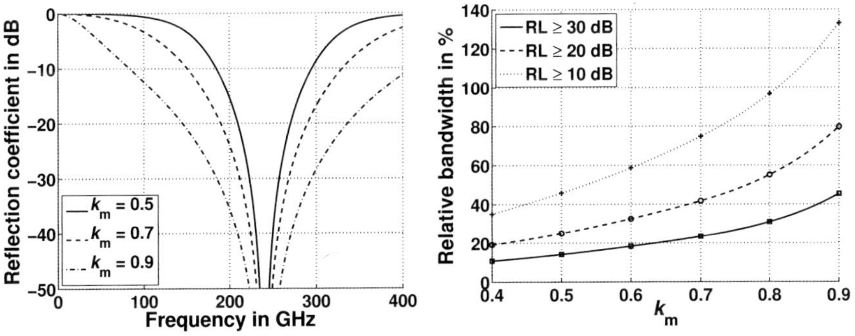

(a) Reflection coefficient versus frequency.

(b) Relative bandwidth versus k_{m}.

Figure 2.8: Bandwidth comparison for transformer unit cells optimized for a maximally flat insertion loss. The curves are obtained by evaluating the ECM depicted in Fig. 2.7.

The bandwidth strongly depends on the coupling coefficient. As the coupling coefficient increases so does the relative bandwidth. For $k_{\mathrm{m}} = 1$ the relative bandwidth reaches 200 %, which is the maximum value any relative bandwidth can achieve.

Realistic coupling coefficients for integrated transformers as part of a MMIC range from 0.6 to 0.9, depending on the available geometries. Since the permeability of the transformer's core material (air, semiconductor material) is one or close to one, a high coupling coefficient has to be achieved by reducing the distance between the primary and secondary inductor and thereby increasing the magnetic flux through both inductors.

From Fig. 2.8(b) it can be seen, that even though realistic coupling coefficients for transformers on MMICs do not exceed 0.9, relative bandwidths (RL $\geq$ 10 dB) of more than 100 % can be realized.

If the response of the DT unit cell does not have to be maximally flat, but is allowed to have a certain ripple, the achievable bandwidth can be even slightly higher. To investigate the possible improvements, values of L and C, that result in a RL better than 10 dB, 20 dB and 30 dB will be determined in the paragraph ahead.

2.3.2 Equal ripple (Chebyshev) frequency response

For a lossless network, the P_{LR} equals the IL (Eqn. 2.9) and an IL ripple of e.g. 0.0436 dB directly results in a RL that has a local minimum of 20 dB at the same frequency ($|IL|^{-1} = 1 - |RL|^{-1}$). To design a DT unit cell (2. order filter structure), which has a predefined ripple and is symmetrical with respect to the design frequency, the maximum of the IL ripple and the local minimum of the RL have to occur directly at the center, i.e. design frequency. Hence, the following conditions have to be fulfilled for a DT design with an equal ripple, i.e. Chebyshev response:

$$P_{\mathrm{LR}}(\omega_0) \stackrel{!}{=} \frac{1}{1 - |\Gamma(\omega_0)|^2}; \quad \left.\frac{\mathrm{d}P_{\mathrm{LR}}}{\mathrm{d}\omega}\right|_{\omega=\omega_0} \stackrel{!}{=} 0; \quad \left.\frac{\mathrm{d}^2 P_{\mathrm{LR}}}{\mathrm{d}^2\omega}\right|_{\omega=\omega_0} \stackrel{!}{\leq} 0 \qquad (2.15)$$

If the P_{LR} of a single DT unit cell (Eqn. 2.10) is inserted in these equations and inequalities, L and C in terms of k_{m}, R_{L} and ω_0 can be determined. Since the derived equations are too complex to be presented as reasonably long equations, the results are given in form of two normalized diagrams versus k_{m} (Fig. 2.9(a), 2.9(b)), which can be used to derive the required inductance (L) and capacitance (C), respectively.

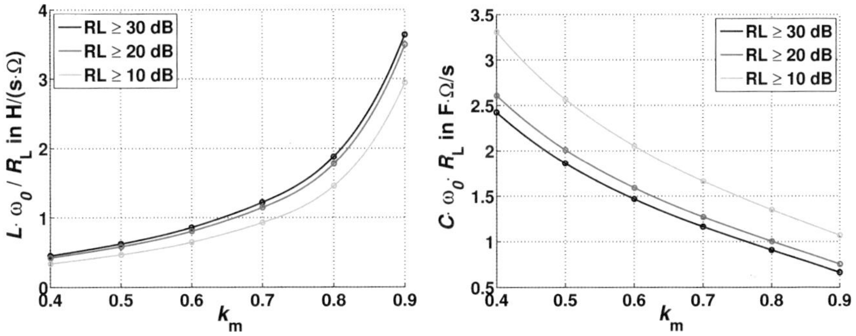

(a) Normalized inductance versus k_{m}.

(b) Normalized capacitance versus k_{m}.

Figure 2.9: Normalized inductance and capacitance for a symmetrical bandwidth with an equal ripple IL i.e. RL $\geq$ 10 dB, 20 dB and 30 dB.

Fig. 2.10(a) depicts the reflection coefficient versus frequency for three different coupling coefficients ($k_{\mathrm{m}} = 0.5$, 0.7 and 0.9) and also three different RL optimizations (RL $\geq$ 10 dB, 20 dB and 30 dB). Here, the unit cells were again designed for a center frequency of 240 GHz.

As already seen for the reflection coefficient of the maximally flat response optimization, the coupling factor has a strong impact on the BW of a single DT unit cell. Its influence is also illustrated in Fig. 2.10(b), where the relative bandwidth versus k_{m} for a return loss better than 10 dB, 20 dB and 30 dB is depicted. Note that this time, the relative BW is extracted from an optimization for exactly the required return loss. This means that for instance the bandwidth for a RL of better than 10 dB has been extracted from an optimization that leads to a RL of exactly 10 dB at the center, i.e. design frequency.

Unlike k_{m}, the optimization for various RL levels at the optimization frequency does only have a small influence on the relative bandwidth. By comparing Fig. 2.10(b) with Fig. 2.8(b), it can be concluded that an optimization regarding the symmetrical bandwidth of a DT unit cell should always focus on the coupling coefficient (k_m) first. If a return loss of better than 20 dB is the objective, a small improvement can be achieved by an optimization that allows for an appropriate RL. For a return loss of less than 20 dB, the benefit of such an optimization is almost negligible in terms of bandwidth. Thus, it can be concluded, that in general an optimization of the DT unit cell for a maximal symmetrical bandwidth can almost always be done by optimizing for a maximally flat or Butterworth response.

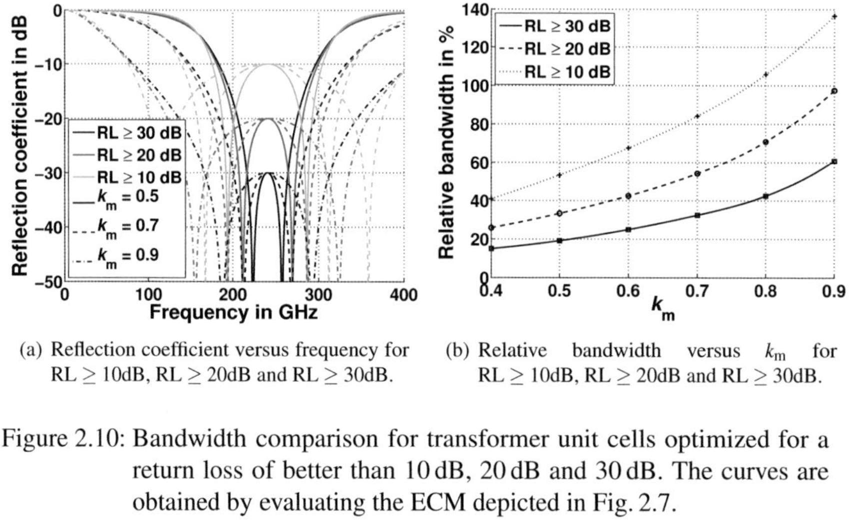

(a) Reflection coefficient versus frequency for RL $\geq$ 10dB, RL $\geq$ 20dB and RL $\geq$ 30dB.

(b) Relative bandwidth versus k_m for RL $\geq$ 10dB, RL $\geq$ 20dB and RL $\geq$ 30dB.

Figure 2.10: Bandwidth comparison for transformer unit cells optimized for a return loss of better than 10 dB, 20 dB and 30 dB. The curves are obtained by evaluating the ECM depicted in Fig. 2.7.

2.3.3 The influence of loss on the response of a DT unit cell optimized for maximum symmetrical bandwidth

To get an idea about the influence of loss on the behavior of a single DT unit cell optimized for a maximum symmetrical bandwidth, Fig. 2.11 demonstrates the deviation of the reflection coefficient ($\Gamma(\omega)$), efficiency (η) and transmission coefficient for various transformer quality factors ($Q = Q_\mathrm{p} = Q_\mathrm{s}$). All transformer responses belong to an optimization for a maximally flat frequency response at 240 GHz.

As expected, the perfectly symmetrical response of a DT unit cell starts to deviate under the influence of transformer loss, which is neglected in the design formulas. However, by inspecting Fig. 2.11(b) and Fig. 2.11(d), it can be seen that despite the loss, the response stays almost symmetrical around the design frequency of 240 GHz, even for very low quality factors of less than 20.

In the passband the influence of the transformer loss on the response of the DT unit cell leads to a slope, which drops from the start of the passband towards its end. The lower the quality factor, the steeper the slope. If the coupling coefficient increases, the bandwidth of the passband increases as well. While the bandwidth increases symmetrical around the design frequency, the frequency at which the DT unit cell shows its highest efficiency decreases. In general, it is always close to the first resonance frequency, which on the other hand means that the influence of loss, especially on the IL at the upper edge of the passband, gets more severe as the coupling coefficient and therefore the bandwidth increases.

In Fig 2.12 the behavior of a DT unit cell for a fixed Q of 20 and various values of k_m can be seen.

Note that the drop of the transmission coefficient is mainly caused by internal loss in the transformer and not by any mismatch that arises due to the neglect of the loss during the derivation of the design formulas. Even though, there is mismatch that increases as loss does, the RL stays well above a level that could cause the depicted loss.

That the mismatch plays only a minor role can also be seen by looking at the efficiency plot in Fig. 2.12(c), which, within the passband region, is only sightly better than the transmission coefficient (Fig. 2.12(d)), even though the efficiency compensates any mismatch at the transformer input ($\eta = \frac{|IL|^{-1}}{1-|RL|^{-1}}$).

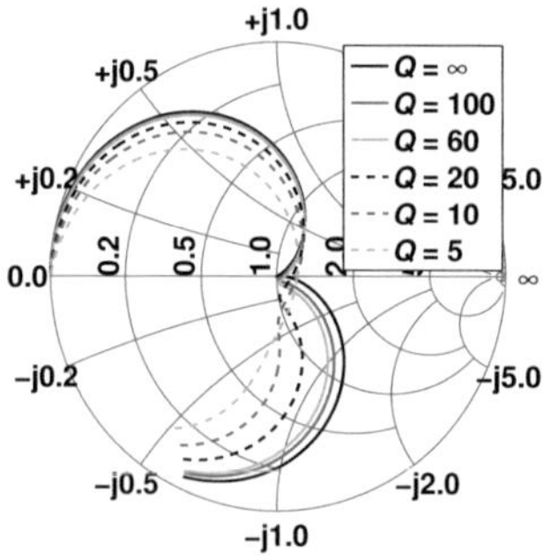

(a) Reflection coefficient

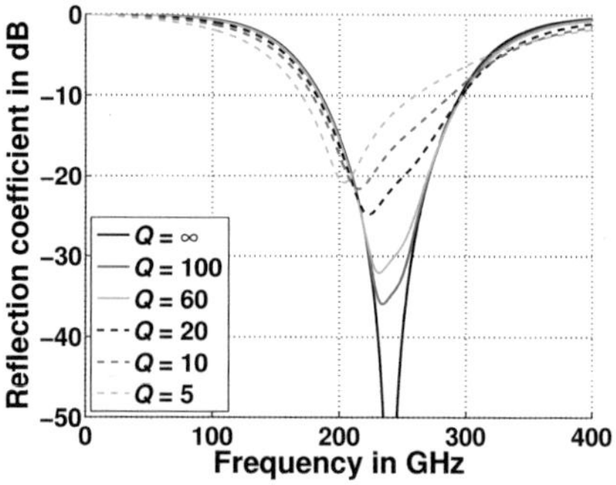

(b) Reflection coefficient

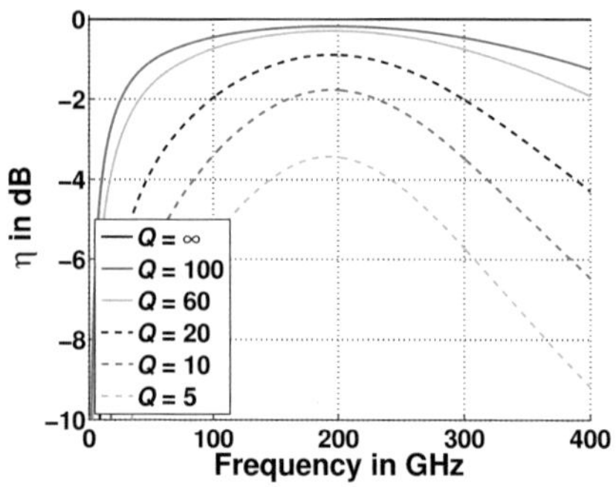

(c) Efficiency

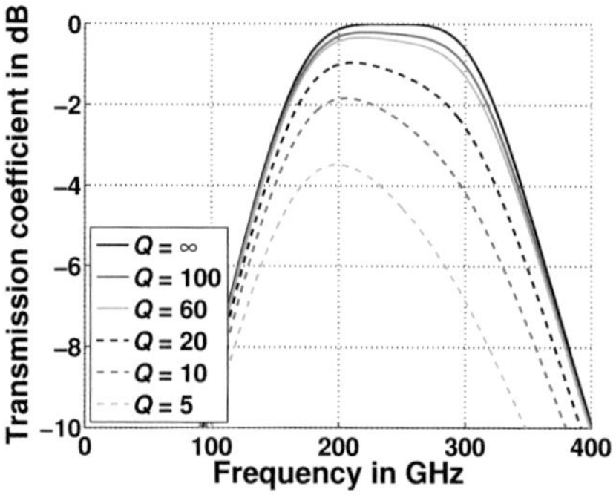

(d) Transmission coefficient

Figure 2.11: Influence of loss ($Q = Q_\mathrm{p} = Q_\mathrm{s}$) on the performance of a 1:1 transformer synthesized for maximally flat response at 240 GHz ($k_\mathrm{m} = 0.5$). The curves are obtained by evaluating the ECM depicted in Fig. 2.3(b), which is terminated with the same load impedance on both sides.

In general, it can be said that the transformer loss degrades both: the return and insertion loss. While the influence on the RL, although it is visible, can usually be tolerated even for very small quality factors, the influence on the IL is severe. Especially DT unit cells with a high k_m and therefore a large bandwidth suffer a notable degradation at the upper edge of their passband. Nevertheless, the possibility to design broadband DTs with a symmetrical bandwidth around the design frequency was proven in this paragraph and will be further investigated in the sections to come.

2.3.4 General conclusion of the lumped DT unit cell analysis

In the previous subsection two optimization strategies were presented.
The first was the standard efficiency optimization based on a lumped ECM, which is the most common approach in literature. By calculating the optimal values of L and C for the DT in terms of efficiency, the lowest possible insertion loss for the design frequency was achieved. On the other hand, it was seen that this approach leads to an unsymmetrical bandwidth around the design frequency. Depending on k_m and Q, the IL and the RL have various ripples, which limit the available bandwidth.
By using more or less the same ECM as for the optimization for maximum efficiency, but applying coupled resonator theory in conjunction with the insertion loss method known from filter theory, the second optimization strategy allows for the widest possible symmetrical bandwidth around the design frequency. Besides equations that lead to a maximum flat DT unit cell response, a design strategy for a RL of better than 10 dB, 20 dB and 30 dB has been presented as well.
The various responses of the different design approaches are compared in Fig. 2.13 for a DT unit cell that is designed for 240 GHz. The quality factor and coupling coefficient of the transformer are $Q = 20$ and $k_m = 0.7$.
As expected, the optimization for maximum efficiency is the only one that results in a perfect match at 240 GHz. The various types of bandwidth optimizations however, all show a wider symmetrical bandwidth around the design frequency.

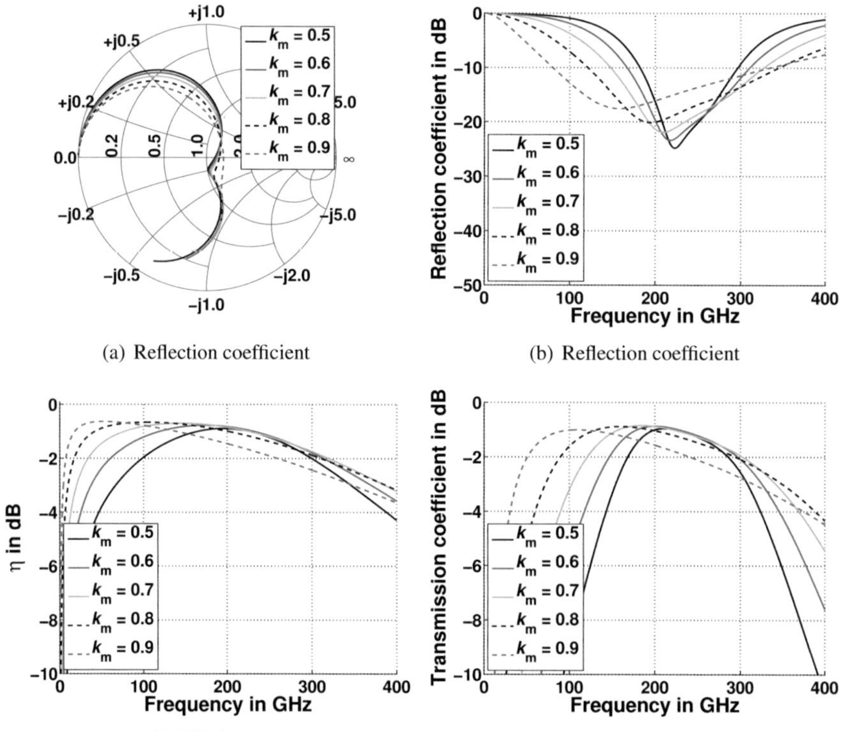

(a) Reflection coefficient

(b) Reflection coefficient

(c) Efficiency

(d) Transmission coefficient

Figure 2.12: Influence of loss ($Q = Q_p = Q_s = 20$) on the performance of a 1:1 transformer synthesized for a maximally flat response at 240 GHz and various coupling coefficients k_m. The curves are obtained by evaluating the ECM depicted in Fig. 2.3(b), which is terminated with the same load impedance on both sides.

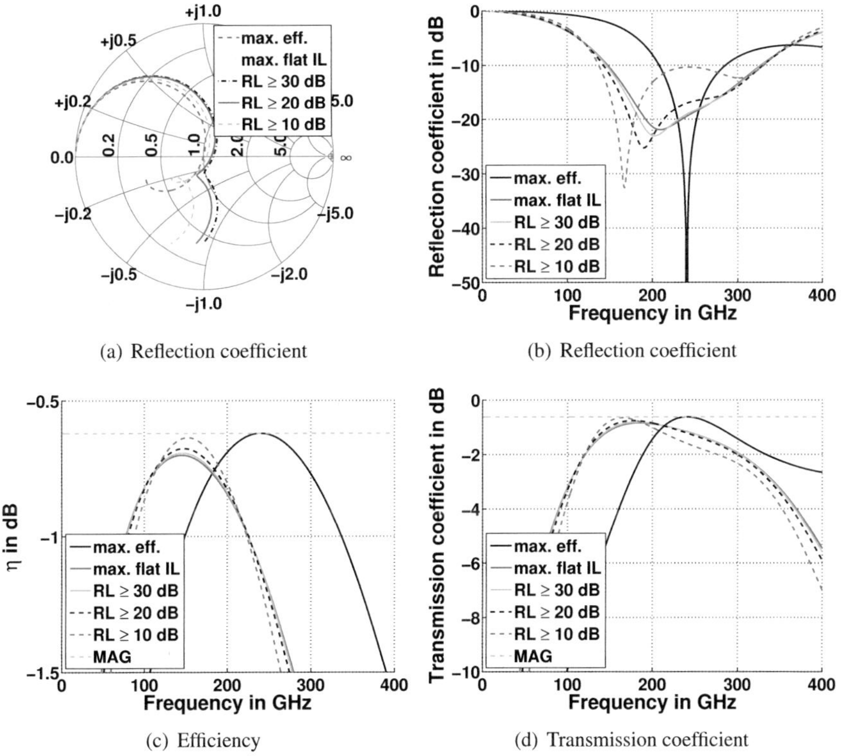

Figure 2.13: Syntheses results of a 1:1 transformer with a $Q_{\mathrm{p}} = Q_{\mathrm{s}} = 20$ and $k_{\mathrm{m}} = 0.7$ optimized for maximum efficiency, maximally flat IL, RL $\geq$ 10dB, RL $\geq$ 20dB and RL $\geq$ 30dB at 240 GHz. The curves are obtained by evaluating the ECM depicted in Fig. 2.3(b), which is terminated with the same load impedance on both sides.

Note that besides the optimization for a return loss of better than 10 dB, non of the bandwidth optimizations in this example achieves the desired RL at the design frequency. This is due to the quite high transformer loss of $Q = 20$, which has been neglected in the design and now alters the matching.
The DT unit cells optimized for a symmetrical bandwidth all have their peak efficiency at about 150 GHz, which, as mentioned before, is close to their first resonance frequency.
Based on the lumped ECM, which is well known from literature, the design procedure for the maximum efficiency at the design frequency has been revised and the maximal available efficiency with respect to Q and k_{m} has been assessed. To evaluate the symmetrical bandwidth that can be achieved for a DT unit cell, a novel design procedure based on coupled resonator and filter theory has been proposed. Like the optimization for maximal efficiency, this approach was also assessed based on a lumped ECM. All the investigations so far have neglected any distributed behavior of the DT unit cell. However, even though the transformer unit cell is small compared to its wavelength, its distributed character cannot be ignored completely, as will be shown in the next section.

2.4 Verification of the bandwidth optimization based on the distributed coupled line unit cell model

To incorporate the effects that are related to the distributed character of the transformer, the lumped coupled inductors of the ECM of the DT unit cell (Fig. 2.2(b)) can be replaced by a pair of coupled lines, which is shorted on one side. The ECM for the new distributed unit cell is depicted in Fig. 2.14. Fig. 2.14(a) shows a semi-distributed and Fig. 2.14(b) a distributed ECM. The difference between both figures is that the semi-distributed model still applies lumped resistors in conjunction with lossless coupled lines to model the DT unit cell, while the distributed model also distributes the loss by using lossy coupled lines. In both ECMs, the inductance of the distributed transformers, i.e. the shorted coupled lines, is compensated by lumped capacitors.

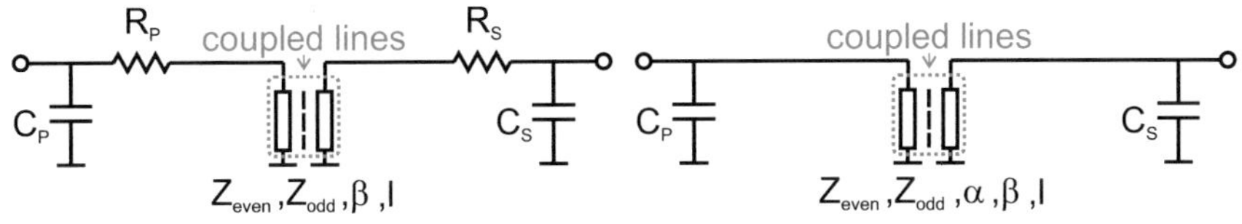

(a) Schematic of a semi-distributed magnetic 1:1 transformer.

(b) Schematic of a distributed magnetic 1:1 transformer.

Figure 2.14: Semi-distributed and distributed model of a single transformer unit cell.

In the previous section, the DT unit cell was completely modeled by ideal lumped elements. Therefore, the inductance of the transformer was frequency invariant. Now that the inductors are modeled by shorted coupled lines, the inductance of the transformer is not frequency invariant anymore, but depends on the characteristic impedance of the two propagating modes (even and odd mode) and their propagation velocity as well as the length of the coupled line. The optimization for a maximum symmetrical bandwidth for the distributed ECM is still performed on the basis of the Eqn. 2.13 and 2.14 as well as the Fig. 2.9(a) and 2.9(b), which were determined by the lumped ECM. For a single design frequency the parameters calculated by the lumped design equations can be converted into the parameters that define a coupled line. For a symmetric coupled line, the following relation between the coupling coefficient of a transformer and the characteristic impedance of the two modes of the coupled line exists [Poz04]:

$$
\begin{aligned}
Z_\mathrm{e} &= Z_\mathrm{k0}\sqrt{\frac{1+k_\mathrm{m}}{1-k_\mathrm{m}}} \\
Z_\mathrm{o} &= Z_\mathrm{k0}\sqrt{\frac{1-k_\mathrm{m}}{1+k_\mathrm{m}}}
\end{aligned}
\tag{2.16}
$$

where Z_k0 is a scaling factor that allows to change the absolute value of the characteristic impedances of the two coupled line modes without altering the coupling coefficient.

The relation between the inductance (L) seen at each port of the symmetric, transverse electromagnetic (TEM) coupled line and its length is as follows:

$$L = \frac{\mathrm{Im}\left\{\frac{Z_o+Z_e}{2} \cdot \tanh\left[(\alpha + j\beta)\, l\right]\right\}}{\omega} = \frac{\mathrm{Im}\left\{\frac{Z_{k0}}{\sqrt{1-k_m^2}} \cdot \tanh\left[(\alpha + j\beta)\, l\right]\right\}}{\omega} \tag{2.17}$$

If the coupled line is lossless or if the loss (α) is small, Eqn. 2.17 can be solved for the line length l, which leads to:

$$l \approx \frac{\arctan(\omega L \frac{2}{Z_e+Z_o})}{\beta} = \frac{\arctan(\omega L \frac{\sqrt{1-k_m^2}}{Z_{k0}})}{\beta} \tag{2.18}$$

To verify that the design equations based on the lumped ECM, which calculate L and C in terms of k_m and R_L, are still valid for a distributed coupled line section instead of coupled lumped inductors, the ECMs in Fig. 2.14 are evaluated next. In this example an optimization for RL = 20 dB at f_c = 240 GHz is performed. The relevant design properties are listed in table 2.1.

f_0	R_L	k_m	$Q_{p,s}$	Z_{k0}	$\varepsilon_{re,o}$	$l_{semi\_dist.}$	$l_{dist.}$
240 GHz	12.5 Ω	0.8	10	50 Ω	2	36.6 µm (4.1 % λ_{eff})	36.6 µm (4.1 % λ_{eff})

Table 2.1: Parameters of the coupled line section used as 1:1 transformer.

As mentioned before, one specific length (l) of the coupled line leads to the port inductance calculated by the design equations for a RL of 20 dB. To verify the correct inductance at the design frequency and to check the variation of the inductance versus frequency, Fig. 2.15(a) shows the self (L) and mutual (M) inductances of the lumped, the semi-distributed and the distributed ECM versus frequency. The inductances of the semi-distributed and distributed ECM, plotted in Fig. 2.15(a), are identical and calculated by the following equations [CdNS+07] from the Z-parameters of the coupled line:

$$L = \frac{\mathrm{Im}\{Z_{11}\}}{\omega} = \frac{\mathrm{Im}\{Z_{22}\}}{\omega}; \qquad M = \frac{\mathrm{Im}\{Z_{12}\}}{\omega} \tag{2.19}$$

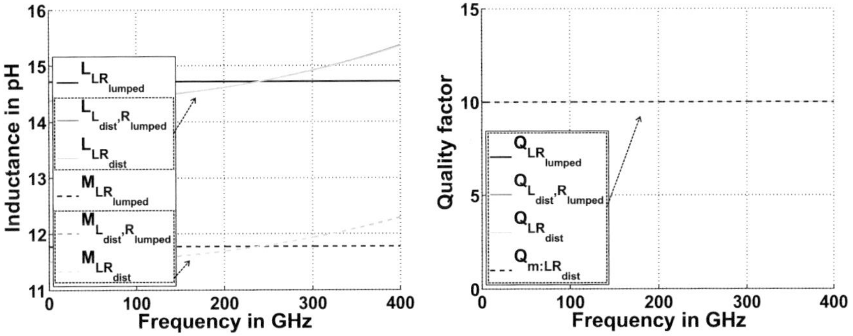

(a) Inductance

(b) Quality factor

Figure 2.15: Comparison between the inductance and quality factor of a lumped, semi-distributed and distributed 1:1 transformer synthesized for maximum 20 dB return loss bandwidth around 240 GHz with a Q_p and Q_s of 10 and $k_\mathrm{m} = 0.8$. The curves are obtained by evaluating the ECMs depicted in Fig. 2.3(b) and 2.14, which are terminated with the same load impedance on both sides.

As expected from Eqn. 2.17, it can be seen that unlike for the lumped model, there is an inductance variation versus frequency. However, at the design frequency the inductances of all three different DT unit cell models are identical.

In Fig 2.15(b), the quality factors of the lumped, the semi-distributed and the distributed ECM versus frequency are potted. For comparison reasons, all quality factors have been set to be 10 over the entire frequency range. To achieve a constant quality factor over frequency the lumped resistors in the lumped and the semi-distributed ECM and α in the distributed ECM have to be varied with frequency. For the semi-distributed and distributed ECM also the variation of the inductance versus frequency has to be taken into account. The quality factors plotted in Fig. 2.15(b) are calculated by the following equations [EGKB07]:

$$Q = \frac{\mathrm{Im}\{Z_{11}\}}{\mathrm{Re}\{Z_{11}\}} = \frac{\mathrm{Im}\{Z_{22}\}}{\mathrm{Re}\{Z_{22}\}}; \qquad Q_{\mathrm{m}} = \frac{\mathrm{Im}\{Z_{12}\}}{\mathrm{Re}\{Z_{12}\}} \tag{2.20}$$

which were applied on the Z-parameters of the two ports of the ECMs of Fig. 2.14 with C_{p} and C_{s} being zero. As demanded the calculated quality factors of the single transformer inductances are 10 for all three different ECMs over the entire frequency range.

Furthermore, in the distributed ECM the mutual quality factor, which is defined as follows

$$Q_{\mathrm{m}} = \frac{\omega M}{R_{\mathrm{m}}} = \frac{\mathrm{Im}\{Z_{12}\}}{\mathrm{Re}\{Z_{12}\}} \tag{2.21}$$

turns out to be exactly the same as the quality factors (Q) of the single transformer inductances of the symmetrical coupled line. For the lumped and semi-distributed ECM the value of Q_{m} is ∞ since $R_{\mathrm{m}} = \mathrm{Re}\{Z_{12}\} = 0$.

In Fig. 2.16 the behavior of the various ECMs (lumped, semi-distributed and distributed) for a DT unit cell is compared.

It becomes obvious, that the lumped and semi-distributed ECMs both show almost identical results. Their reflection and transmission coefficients would lead to the assumption that a DT unit cell with quality factors of $Q = 10$ is not suitable for being used for DTs optimized for maximum symmetrical bandwidth and a RL of 20 dB at the optimization frequency. On the other hand, the response for the distributed ECM, even though it also possesses Q values of only 10, is obviously capable of delivering the necessary performance. Its efficiency is decisively higher than that of the other two. The reason for this is the resistive coupling between the two coupled lines, as will be shown in the section ahead. Due to R_{m} not being 0 or close to it, the efficiency of the overall transformer is improved.

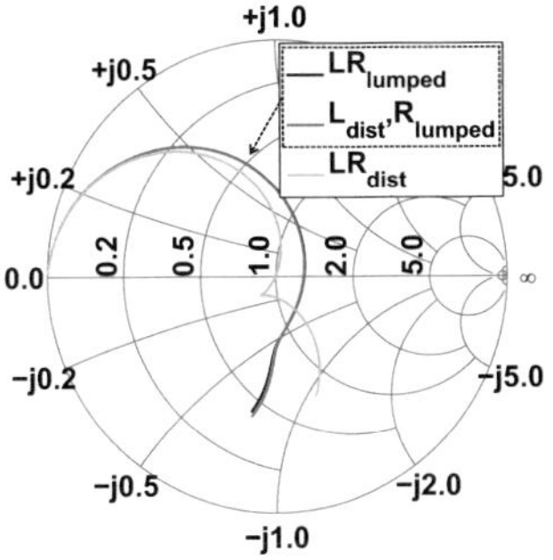

(a) Reflection coefficient

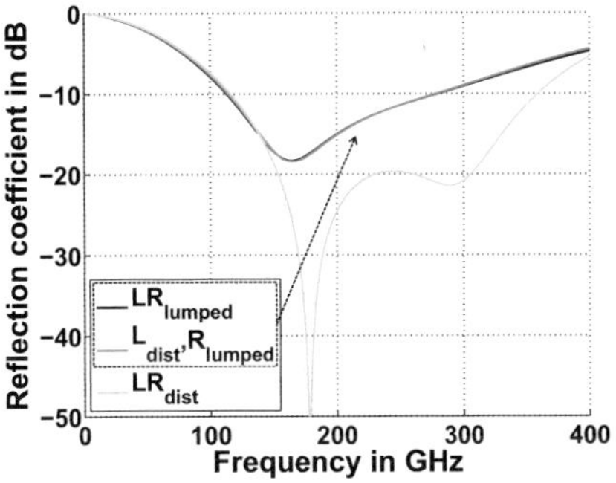

(b) Reflection coefficient

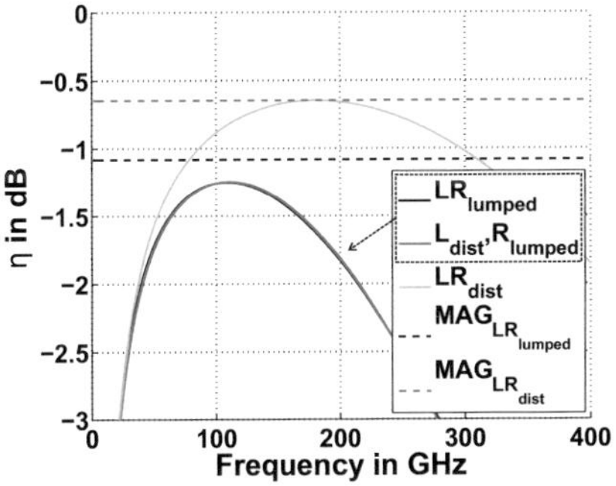

(c) Efficiency

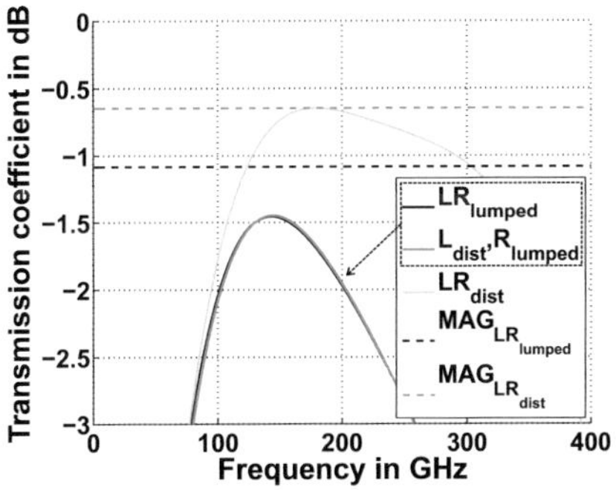

(d) Transmission coefficient

Figure 2.16: Comparison between the lumped, semi-distributed and distributed 1:1 transformer synthesized for max. 20 dB return loss bandwidth around 240 GHz with a Q of 10 and $k_\mathrm{m} = 0.8$. The curves are obtained by evaluating the ECMs depicted in Fig. 2.3(b) and 2.14, which are terminated with the same load impedance on both sides.

2.5 Efficiency optimization based on the distributed coupled line unit cell model

In the previous section, it was shown that the bandwidth optimization method, which is based on equations derived from a lumped ECM, can also be applied, if the transformer is not modeled by coupled lumped inductors but a pair of coupled transmission lines. By comparing the results of the optimization for a lumped and a distributed ECM of a DT unit cell, it was shown that the distributed ECM has a better performance in terms of efficiency.
Therefore, in this section, a similar analysis will be performed for an optimization for maximum efficiency at a given frequency. Based on the standard design procedure and equations, which have been introduced in section 2.2, the lumped, semi-distributed and the distributed ECM will be compared again.
Table 2.2 shows the relevant parameters of the ECMs, which are either set or determined by the optimization equations derived from the lumped model introduced at the beginning of this chapter.

f_0	R_L	k_m	$Q_{p,s}$	Z_{k0}	$\varepsilon_{re,o}$	$l_{semi\_dist.}$	$l_{dist.}$
240 GHz	12.5 Ω	0.8	10	50 Ω	2	10.3 µm ($1.2\,\%\,\lambda_{eff}$)	10.3 µm ($1.2\,\%\,\lambda_{eff}$)

Table 2.2: Parameters of the coupled line section used as 1:1 transformer.

The corresponding response of the port inductance versus frequency is depicted in Fig. 2.17 for the various models. Due to the smaller length of the coupled line pair, the inductance of the semi-distributed and distributed model does not vary as much as for the optimization for maximum symmetrical bandwidth and 20 dB RL. The relation between the inductance and the length is as defined in Eqn. 2.17. Again, the quality factor of all ECMs is fixed to be 10 and the values of R_p, R_s and α have been determined to fulfill this requirement.
The results of the efficiency optimization applied on the various ECMs are visualized in Fig. 2.18.

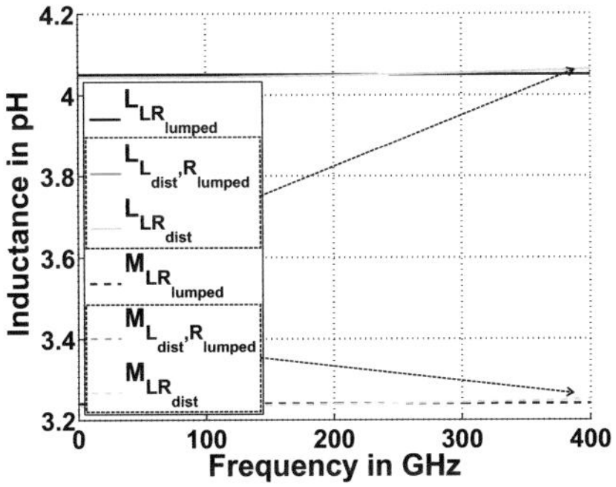

Figure 2.17: Comparison between the inductance of a lumped, semi-distributed and distributed 1:1 transformer synthesized for optimum efficiency at 240 GHz with a Q of 10 and $k_{\mathrm{m}} = 0.8$. The curves are obtained by evaluating the ECMs depicted in Fig. 2.3(b) and 2.14, which are terminated with the same load impedance on both sides. The inductance at 240 GHz was determined by applying the standard optimum efficiency equation model (Eqn. 2.1 - 2.7).

In these plots it becomes obvious that the distributed transformer model, based on coupled lines, can no longer be represented by the original lumped ECM revised in section 2.2 and widely used in literature for the design of DTs and DATs. As visible in the Fig. 2.18(c) and 2.18(d), the optimization for the distributed unit cell, unlike those for the lumped and semi-distributed unit cells, does not result in the highest possible efficiency at the design frequency of 240 GHz. Its peak is higher than the maximum possible efficiency for a transformer, based on the lumped ECM and it lies at a higher frequency.

As mentioned before, the reason why the response of the distributed ECM deviates from that of the lumped and semi-distributed ECM is the resistive coupling that occurs in lossy coupled lines. The resistive coupling is not modeled in the standard DT/DAT ECM, which leads to a wrong efficiency calculation and in the end a wrong efficiency optimization, i.e. wrong values for L and C_s [CHH+10, HCH12]. That resistive coupling plays a significant role for the behavior of an integrated transformer has been discussed in multiple publications [NRB02, SYK+07, WLX+09, BPKL10], but its effect has never been accounted for during any transformer optimization so far. The main reason for the resistive coupling are hybrid effects of parasitic capacitances and eddy currents in the substrate.

To determine the efficiency of a transformer, including the effects of resistive coupling, its S-parameters can be evaluated as indicated in Eqn. 2.22:

$$\eta = \frac{|S_{21}|^2}{1 - |S_{11}|^2} \tag{2.22}$$

If the lumped ECM of a DT models the real distributed transformer correctly, the efficiency calculated by the ECM and that from the S-parameters (Eqn. 2.22) have to be identical. Since the distributed and the lumped ECM deviate from each other, this means, that the lumped ECM in its current state, cannot be applied to model a DT unit cell, if resistive coupling plays a significant role ($R_m \not\approx 0$).

To find the difference between the coupled line and the transformer T-model applied in the lumped ECM, the ECM of a coupled line will be investigated next. In [KH70, MBBH07] a distributed equivalent unit cell circuit of two uniformly coupled lossless asymmetrical transmission lines is proposed (Fig. 2.19). It is basically composed of two standard lumped element circuit models for a single lossless transmission line [Poz04] and extended by an additional parallel capac-

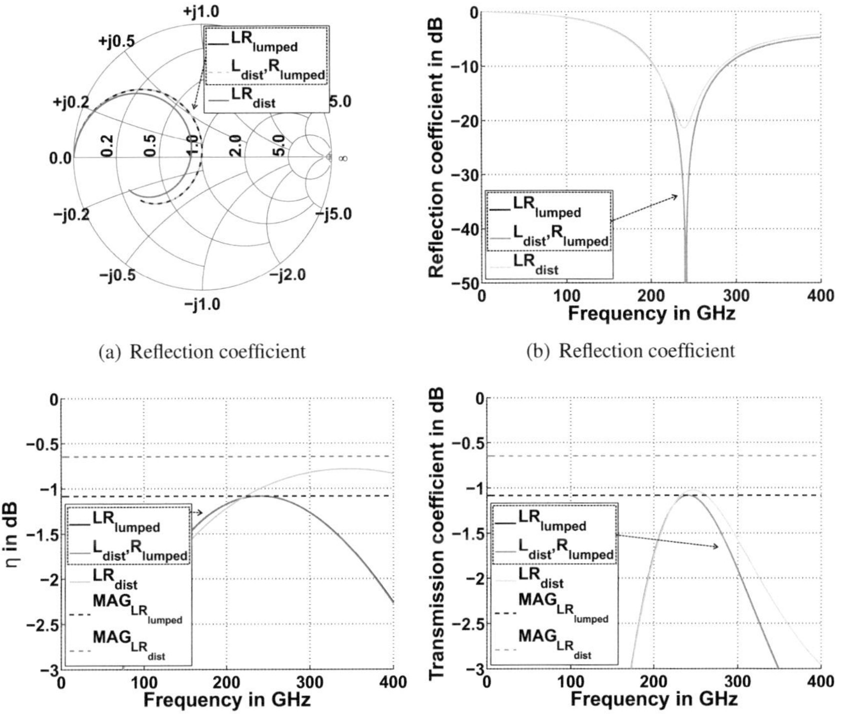

(a) Reflection coefficient

(b) Reflection coefficient

(c) Efficiency

(d) Transmission coefficient

Figure 2.18: Comparison between the lumped, semi-distributed and distributed 1:1 transformer synthesized for optimal efficiency at 240 GHz with a Q of 10 and $k_\mathrm{m} = 0.8$. For these figures the standard efficiency optimization was applied (Eqn. 2.6 - 2.7). The curves are obtained by evaluating the ECMs depicted in Fig. 2.3(b) and 2.14, which are terminated with the same load impedance on both sides.

itor and the mutual inductance between the series inductors, to account for the capacitive and inductive mutual coupling.

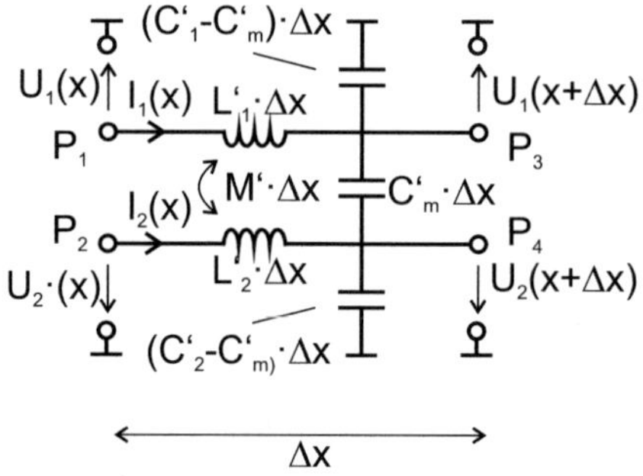

Figure 2.19: Lossless coupled line ECM.

The voltages and currents on the coupled lines are governed by the following differential equations

$$\begin{aligned} -\frac{dU_1}{dx} &= z_1 I_1 + z_m I_2 \\ -\frac{dU_2}{dx} &= z_2 I_2 + z_m I_1 \\ -\frac{dI_1}{dx} &= y_1 U_1 + y_m U_2 \\ -\frac{dI_2}{dx} &= y_2 U_2 + y_m U_1 \end{aligned} \tag{2.23}$$

where for a lossless coupled line the inductance and admittance values are:

$$\begin{aligned}
z_1 &= j\omega L_1' \\
z_2 &= j\omega L_2' \\
z_{\mathrm{m}} &= j\omega M' \\
y_1 &= j\omega C_1' \\
y_2 &= j\omega C_2' \\
y_{\mathrm{m}} &= -j\omega C_{\mathrm{m}}'
\end{aligned} \tag{2.24}$$

The line parameters and the S-parameters of a lossless coupled line can be calculated by using the differential equations given in 2.23. The procedure is closely related to the calculation of line parameters from the telegrapher's equation [Poz04] and can be found for instance in [MBBH07]. If the S-parameters of a coupled line are given, the reverse calculation is also possible. The S-parameters can be used to extract the per unit-length quantities (R′,L′,G′,C′). This can be done in a similar way as for the telegrapher's equation model of a standard two port (quasi) TEM transmission line [EE92, Zwi05]. The S-parameters of a 2N-port network can be converted in ABCD-parameters, as proposed in [LCJ$^+$03] and for the sake of completeness revised in appendix C.

If the ABCD-parameters are available, these parameters can be used to calculate the per unit length quantities by means of an eigenvalue decomposition [MS00] and multiple other matrix operations. Since this procedure is quite complex, it will not be repeated here, but the interested reader is referred to [KKOY06, Sam08].

If the coupled line encompasses loss ($\alpha \neq 0$ - Fig. 2.20(a)), the inductance ($z_{1,2,\mathrm{m}}$) and admittance ($y_{1,2,\mathrm{m}}$) parameters in Eqn. 2.23, which are determined from given S-parameters of a line, are not just imaginary, but contain real parts that represent the loss.

Therefore, Eqn 2.24 becomes:

$$
\begin{aligned}
z_1 &= R'_1 + j\omega L'_1 \\
z_2 &= R'_2 + j\omega L'_2 \\
z_\mathrm{m} &= R'_\mathrm{m} + j\omega M' \\
y_1 &= G'_1 + j\omega C'_1 \\
y_2 &= G'_2 + j\omega C'_2 \\
y_\mathrm{m} &= -G'_\mathrm{m} - j\omega C'_\mathrm{m}
\end{aligned}
\tag{2.25}
$$

The lossless ECM shown in Fig. 2.19 can be extended in a way that each real and imaginary part of the impedance and admittance components of the differential equations given in Eqn 2.23 is represented by one lumped element. The extended ECM of a lossy coupled line is depicted in Fig. 2.20(b) [HHKE09].

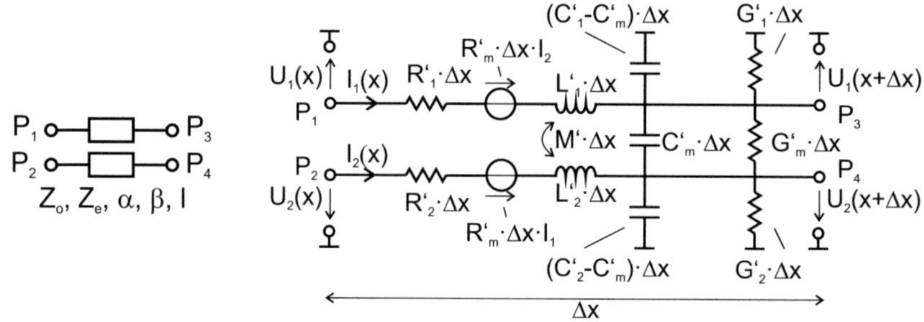

(a) Single coupled line element.

(b) Extended lossy coupled line unit cell ECM.

Figure 2.20: The equivalent lumped lossy coupled line model.

While most of the real parts of the impedance and admittance parameters are represented by lumped resistors, the mutual resistance R'_m is modeled by a current controlled voltage source in the lossy ECM. As determined by the differential equations of the coupled lines (Eqn. 2.23), the voltage sources model the voltage drop on one of the lines in dependency of the current on the other line ($\frac{dU_{1,2}}{dx} = -R'_\mathrm{m} I_{2,1}$).

R'_m represents exactly the mutual resistive coupling, which is neglected in the standard ECM used for the efficiency optimization of DTs. Depending on the loss of the coupled line, R'_m can become significantly high, which means that it actually cannot be neglected, neither in the lumped ECM of a coupled line unit cell nor in the lumped ECM of a DT.

In table 2.2(b) the per unit length quantities extracted from S-parameters of a standard ADS (Agilent's Advanced Design System) coupled line model (Fig. 2.20(a)) are listed. The ADS model implements the standard equations for a symmetrical TEM coupled line [Poz04] and its parameters are listed in table 2.2(a).

(a) Coupled line parameters:

Z_e	150	Ω
Z_o	16.6667	Ω
ε_{r_e}	2	
ε_{r_o}	2	
α_e	7008	dB/m
α_o	7008	dB/m

(b) Lumped ECM parameters (R′L′G′C′):

$R'_{1,2}$	67239	Ω/m
R'_m	53791	Ω/m
$L'_{1,2}$	393.11	nH/m
M'	314.5	nH/m
$G'_{1,2}$	26.89	S/m
G'_m	21.52	S/m
$C'_{1,2}$	157.24	pF/m
C'_m	125.79	pF/m

Table 2.3: Parameters of a coupled line and its extracted per unit length quantities.

Note that in general the loss of a transmission line is frequency dependent, which means that all resistors as well as the current controlled voltage sources have to be frequency dependent as well. Therefore, the resistance and conductance parameters listed in table 2.2(b) are actually only valid for a single frequency. However, for the sake of simplicity and without loss of generality, it will be assumed for now that the loss is frequency independent.

To find out, whether R'_m plays a relevant role for the behavior of the lumped model, the reflection and transmission coefficients of a lossy coupled line and a number of cascaded lumped coupled line unit cells, resembling the same length ($l = 100\,\mu m$),

are compared. In the Fig. 2.21(a) and 2.21(b) the lumped ECM neglects the mutual resistance ($R'_{\mathrm{m}} = 0$), which results in a deviation between both models. However, Fig. 2.21(c) and 2.21(d) show reflection and transmission coefficients of the lumped model, which are indistinguishable from the real distributed coupled line model. In these plots R_{m} was incorporated in the lumped line model.
The curves of the transmission coefficients for the lumped model without R'_{m} implementation show a smaller magnitude than the distributed coupled line model, which means that a lumped ECM of a coupled line predicts too much loss, if R'_{m} is neglected. Note that this is exactly the same behavior, that can be seen for the lumped ECM of the DT.
Hence, if R_{m} is incorporated in the lumped ECM of a DT, it should be possible to get a better agreement between the distributed and the lumped ECM.
If the overall length of the coupled line is small compared to the guided wavelength (λ_{eff}), the effect of distribution becomes also small, which means that the behavior of a coupled line can be sufficiently modeled by a single lumped coupled line unit cell model.
As mentioned before, a single 1:1 transformer of a DT is supposed to be small compared to the guided wave length ($l \ll \lambda_{\mathrm{eff}}$), which is why the lumped coupled line ECM unit cell can be adopted to model a single DT unit cell as depicted in Fig. 2.22(a).
Since the length of the grounded, i.e. shorted coupled line, is small, the voltage on both lines is small as well, which enables further simplification, as all admittance elements can be neglected in the first order. Therefore, the ECM of a shorted coupled line can be rearranged as presented in Fig. 2.22(b).
Note that the ECM shown in Fig. 2.22(b) looks very similar to the ECM of Fig. 2.3(a), which was the origin of the original efficiency optimization equations. If parallel capacitors are added and the series inductance of both coupled lines are assumed to be identical ($L' = L'_1 \approx L'_2$), the only difference are the additional current controlled voltage sources, representing the resistive coupling.
Fig. 2.23(a) depicts the novel lumped ECM, which will be the origin of a new DT efficiency optimization that takes resistive coupling into account. As for the original optimization the first step towards a new, more accurate equation for the efficiency of a DT is to replace the lumped transformer by its equivalent T-model and to add the load. In Fig. 2.23(b) the resulting lumped ECM is shown. To ease

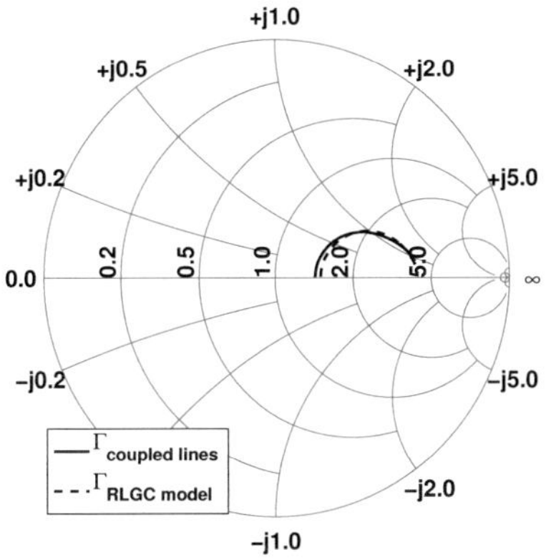

(a) Reflection coefficient ($R'_\mathrm{m} = 0\,\Omega/m$)

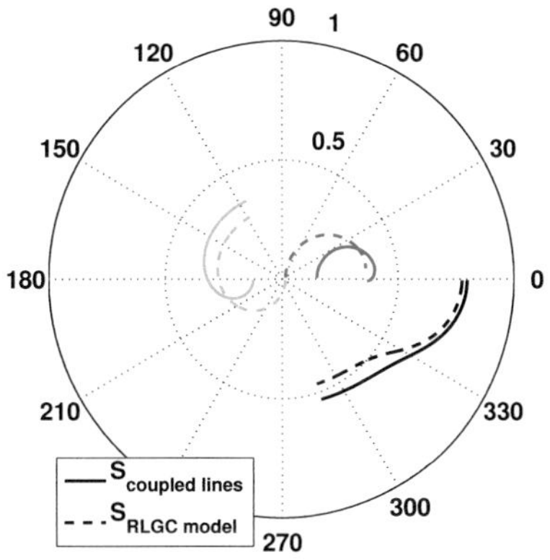

(b) All three transmission coefficients ($R'_\mathrm{m} = 0\,\Omega/m$)

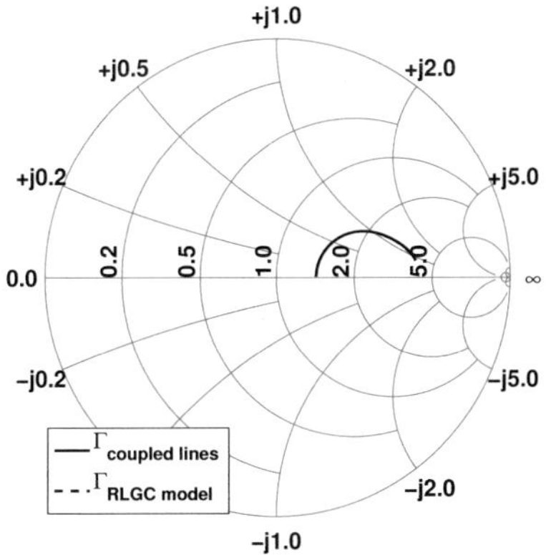

(c) Reflection coefficient ($R'_\mathrm{m} = 53791\,\Omega/m$)

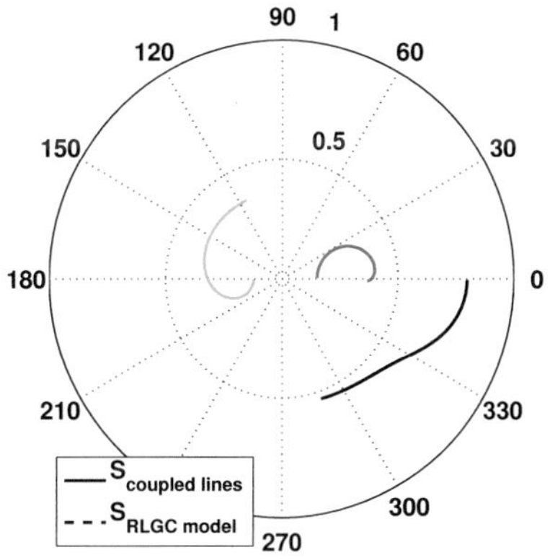

(d) All three transmission coefficients ($R'_\mathrm{m} = 53791\,\Omega/m$)

Figure 2.21: Comparison between a distributed symmetrical coupled line model with frequency invariant α and an R′L′G′C′ model of the symmetrical coupled line with ($R'_\mathrm{m} = 53791\,\Omega/m$) and without ($R'_\mathrm{m} = 0\,\Omega/m$) R'_m implementation. By implementing R'_m correctly, the S-parameters of the R′L′G′C′ model and the distributed coupled line are identical. The simulated coupled line has a length of 100 µm and is evaluated from 0 GHz to 400 GHz.

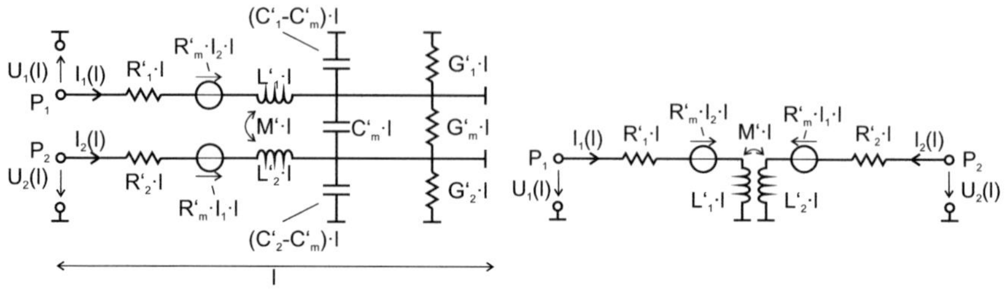

(a) Schematic of a shorted coupled line creating a distributed transformer.

(b) Rearranged schematic of a shorted coupled line resulting in a DT.

Figure 2.22: From the ECM of a lossy coupled line towards a novel extended lumped ECM of a DT.

the derivation of the efficiency, the parallel capacitor (C_S) on the secondary side and the load resistance (R_L) have already been replaced by an equivalent series capacitance (C_{eq} - Eqn. 2.1) and resistance (R_{eq} - Eqn. 2.1).

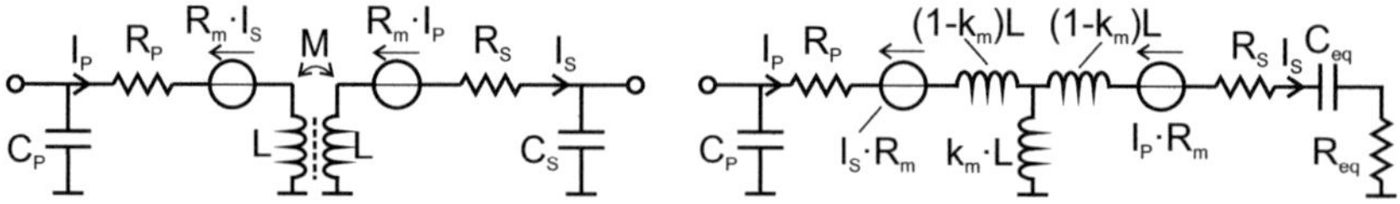

(a) Novel schematic of a DT, which incorporates the resistive coupling.

(b) Novel lumped ECM.

Figure 2.23: The novel extended lumped element model used to derive the efficiency of a DT.

Based on the ECM presented in Fig. 2.23(b) the efficiency of a DT that includes the resistive coupling can be determined. The complete derivation is presented in appendix D and only the final equation of the derivation is shown in Eqn. 2.26 where $Q_m = \frac{\omega M}{R_m} = \frac{\omega k_m L}{R_m}$ and R_{eq}, C_{eq} are frequency dependent (Eqn. 2.1).

$$\eta = \frac{R_{\mathrm{eq}}}{R_{\mathrm{eq}} + \frac{\omega L}{Q_{\mathrm{s}}} + \left| \frac{j\omega L + \frac{\omega L}{Q_{\mathrm{s}}} + R_{\mathrm{eq}} + \frac{1}{j\omega C_{\mathrm{eq}}}}{j\omega k_{\mathrm{m}} L + \frac{\omega k_{\mathrm{m}} L}{Q_{\mathrm{m}}}} \right|^2 \cdot \frac{\omega L}{Q_{\mathrm{p}}} - 2\,\mathrm{Re}\left\{ \frac{j\omega L + \frac{\omega L}{Q_{\mathrm{s}}} + R_{\mathrm{eq}} + \frac{1}{j\omega C_{\mathrm{eq}}}}{j\omega k_{\mathrm{m}} L + \frac{\omega L}{Q_{\mathrm{m}}}} \right\} \cdot \frac{\omega k_{\mathrm{m}} L}{Q_{\mathrm{m}}}} \tag{2.26}$$

To demonstrate the superiority of the new η equation, Fig. 2.24 shows the efficiency of the coupled line based DT from Fig. 2.18 calculated by the old (Eqn. 2.5) and the new (Eqn. 2.26) η equation as well as the efficiency determined from S-parameters (Eqn. 2.22) of the coupled line DT model. Unlike the old equation used in literature, the new equation for the efficiency of a DT mirrors the real efficiency determined from the S-parameters of the transformer up to 400 GHz.

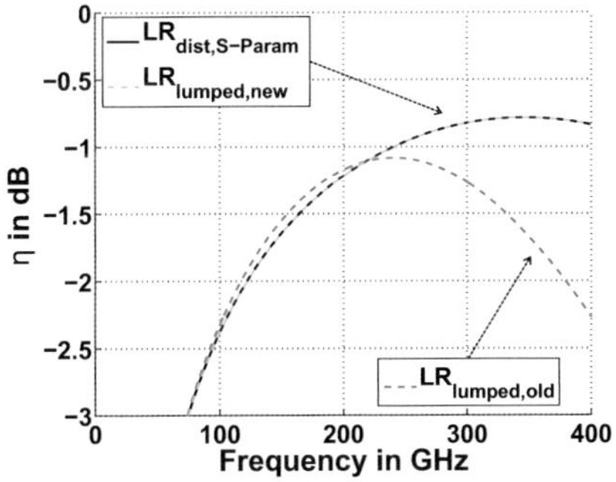

Figure 2.24: Comparison between the old and the new efficiency equations applied for the coupled line based DT investigated previously (Fig. 2.18). The efficiencies are calculated by means of S-parameters obtained from the evaluation of the ECM shown in Fig 2.14(b) ($\mathrm{LR}_{\mathrm{dist,S-Param}}$), the equation derived from the ECM used in literature ($\mathrm{LR}_{\mathrm{lumped,old}}$ - Fig. 2.4) and the new ECM developed in this chapter ($\mathrm{LR}_{\mathrm{lumped,new}}$ - Fig. 2.23).

To determine the optimum values of C_s and L with respect to the quality factors of the transformer inductance on basis of the new η equation (2.26), its partial differentiation with respect to C_s and L can be set to zero and solved for C_s and L respectively (Eqn. 2.27).

$$\frac{\partial \eta}{\partial C_s \partial L} \stackrel{!}{=} [0,0] \tag{2.27}$$

(R_{eq} and C_{eq} have to be replaced as defined in Eqn. 2.1.)
Hereby, the optimum values for C_s and L are determined to be as follows:

$$C_s = -\frac{\left(-Q_m + k_m^2 Q_p\right) Q_s}{\omega_0 \sqrt{Q_m^2 - Q_s^2 k_m^4 Q_p^2 + Q_m^2 Q_s k_m^2 Q_p - Q_s k_m^2 Q_p R_L}} \tag{2.28}$$

$$L = \frac{\sqrt{Q_m^2 - Q_s^2 k_m^4 Q_p^2 + Q_m^2 Q_s k_m^2 Q_p - Q_s k_m^2 Q_p Q_s Q_m R_L}}{\left(Q_s^2 Q_m^2 + Q_m^2 + Q_m^2 Q_s k_m^2 Q_p - 2k_m^2 Q_s^2 Q_p Q_m - Q_s k_m^2 Q_p\right) \omega_0} \tag{2.29}$$

Eqn. 2.28 and 2.29 represent the first equations for the design of DTs/DATs, which incorporate the mutual resistive coupling and therefore lead to an optimum efficiency even if R_m is not negligible.
To verify the proper optimization in terms of efficiency and to visualize the difference between the old and new design equations, a DT unit cell based on a distributed coupled line is optimized with the new design equations and compared to DTs based on the lumped and semi-distributed ECMs designed by the old design equations. The general properties of the DTs are the same and they are listed in table 2.4.

f_0	R_L	k_m	$Q_{p,s}$	Z_{k0}	$\varepsilon_{re,o}$	$l_{semi\_dist.}$	$l_{dist.}$
240 GHz	12.5 Ω	0.8	10	50 Ω	2	10.3 µm (1.2 % λ_{eff})	27.7 µm (3.1 % λ_{eff})

Table 2.4: Parameters of the coupled line section used as 1:1 transformer.

In Fig. 2.25 the inductance for the lumped, the semi-distributed and the distributed ECM after an optimization based on the old and the new design equations is shown.

As can be seen, the optimal inductance calculated by the new equations is more than twice as high as those for the lumped and semi-distributed model. This leads to a length difference between the coupled line applied in the semi-distributed and the distributed ECM, which is quite severe. The actual lengths of the coupled line in the semi-distributed and the distributed ECM, leading to the optimum port inductance, are listed in table 2.4.

The quality factors of the inductive transformers are exactly 10 over the entire frequency range.

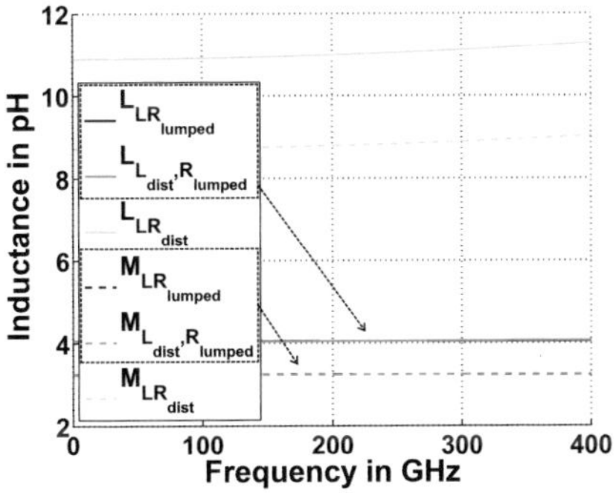

Figure 2.25: Comparison between the inductance of a lumped, semi-distributed and distributed 1:1 transformer synthesized for optimum efficiency at 240 GHz with a Q_p and Q_s of 10 and $k_m = 0.8$. The curves are obtained by evaluating the ECMs depicted in Fig. 2.3(b) and 2.14, which are terminated with the same load impedance on both sides. The inductance of the distributed transformer at 240 GHz is determined by applying the newly developed optimum efficiency equations (2.28 - 2.29).

The results of the optimization can be compared in Fig. 2.26. There, the responses of the lumped and semi-distributed ECM, optimized with the old design equations, are exactly the same as they were in Fig. 2.18, but this time also the distributed ECM designed with the new design equations peaks at 240 GHz. This means that encompassing R_m in the design equations, led to the expected improvement for the

design of DTs, which show a significant amount of mutual resistance. As already seen during the investigation of the distributed ECM for the optimization for optimal symmetrical bandwidth, the mutual resistive coupling improves the efficiency of the transformer, which is why the peak of the efficiency lies higher than that for the lumped and semi-distributed model. (Recall that Q_m of a symmetrical, TEM coupled line equals Q_1 and Q_2.)
Due to the fact that in this example the optimum length of the coupled line for a DT for optimum efficiency at 240 GHz is already quite close to the optimum length of a coupled line applied in a DT optimized for maximal symmetrical bandwidth at the same frequency (table 2.1), the bandwidth of the transformer designed by the new equations is also more broadband than the response of the ideal lumped and semi-distributed model.
Based on the new equations for a lumped ECM, which takes the mutual resistive coupling into account, the maximum efficiency of a DT unit cell can be calculated in terms of the quality factors ($Q = Q_p = Q_s = Q_m$) and the mutual magnetic coupling (k_m). To do so, the equations for the optimum values of C_s (Eqn. 2.28) and L (Eqn. 2.29) have to be entered into Eqn. 2.26. The resulting maximum efficiency is given in Eqn. 2.30.

$$\eta_{max} = \frac{Q_p(Q_m^2+1)Q_s k_m^2}{-Q_s k_m^2 Q_p + Q_m^2 Q_s k_m^2 Q_p + 2\sqrt{Q_m^2 - Q_s^2 k_m^4 Q_p^2 + Q_m^2 Q_s k_m^2 Q_p - Q_s k_m^2 Q_p Q_m + 2Q_m^2}} \tag{2.30}$$

Due to the resistive coupling, the maximum achievable efficiency of a DT unit cell is higher than what was predicted with the old lumped ECM that ignored the resistive coupling. Fig. 2.27 compares the contours of the old and the new ECM versus Q and k_m.
As can be seen, the difference becomes more severe as the coupling coefficient increases and the quality factor drops. From Fig. 2.27 it can be concluded, that in principle the DT concept is capable of providing an acceptable efficiency even if the quality factor is low.

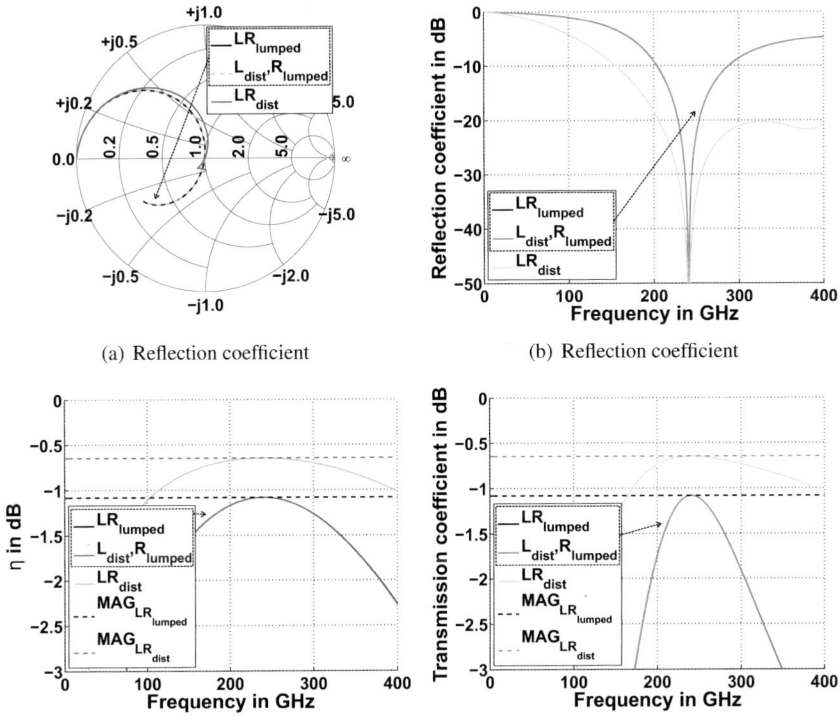

(a) Reflection coefficient

(b) Reflection coefficient

(c) Efficiency

(d) Transmission coefficient

Figure 2.26: Comparison between the old lumped, semi-distributed and distributed 1:1 transformer ECMs synthesized for optimum efficiency at 240 GHz with a Q_p and Q_s of 10 and $k_m = 0.8$. For the curves obtained with the old lumped (Fig. 2.4(a)) and semi-distributed (Fig. 2.14(a)) ECMs the old optimization equations (Eqn. 2.6, 2.7) were used and for the distributed ECM (Fig. 2.14(b)) the newly developed efficiency equations (Eqn. 2.28 - 2.29) were applied. All ECMs are terminated with the same load impedance on both sides.

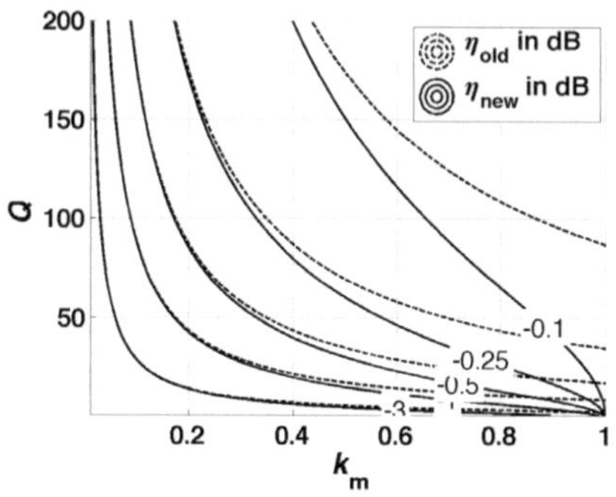

Figure 2.27: Maximum efficiency (η_{max}) versus quality factor (Q) and magnetic coupling coefficient (k_m) calculated on basis of the old (Eqn. 2.8) and new (Eqn. 2.30) maximum efficiency formula.

2.6 From a single unit cell towards a distributed transformer

So far, all presented optimization strategies have been applied on single DT unit cells. In this section, a number of unit cells will be cascaded to verify that the resulting DTs do have the same response as the unit cells they are composed of. It will be shown that as long as the overall DT is small compared to the guided wavelength, the deviation of the DT performance from that of a single unit cell is negligible.

Whether an efficiency or bandwidth optimized unit cell can be considered small compared to the wavelength, depends in general on the optimal inductance value, which is calculated by the optimization equations. The inductance depends mainly on the load termination (R_L) and the coupling coefficient (k_m) but also on the losses within the magnetic transformer (Q).

The inductors of the unit cell are implemented as distributed coupled lines, which means that the optimum inductance is mainly achieved by optimizing the length of a given coupled line. If the required inductance value is high, the length of the coupled line can reach a relative length that cannot be considered small anymore. If that happens, the performance of the optimized DT, consisting of N unit cells, degrades.

Of all the DT optimization methods introduced in this thesis, the optimization for maximally flat response requires the highest and the optimization for optimum efficiency the smallest inductance value. That the optimization for a symmetrical bandwidth returns higher inductance values was already visible in table 2.1 and 2.4, where the higher inductance value of the optimization for maximum symmetrical BW with a 20 dB RL at the optimization frequency results in a longer coupled line compared to the optimization for maximum efficiency.

Within this section, the influence of the parameters of a symmetrical TEM coupled line will be investigated. Based on the optimization for maximally flat IL at 240 GHz, the imbalance in terms of amplitude and phases between the different ports, together with the port matching of the DT, will be compared. As mentioned before, in general the optimization for maximally flat IL results in the largest inductance, i.e. the longest coupled lines. Therefore, the deviation between the performance of a single coupled line and that of a DT composed out of multiple unit cells optimized for maximally flat response will be the highest among all the optimizations compared in this thesis.

In Fig. 2.28 the reflection and transmission coefficients for a three, four and five port DT are presented. The unit cells, which are cascaded to build the DT, all consist of lossless ($Q = \infty$), symmetrical, TEM coupled lines with a coupling coefficient $k_m = 0.5$ and a characteristic impedance scaling factor $Z_{k0} = 50\,\Omega$, i.e. $Z_e = 86.6\,\Omega$ and $Z_o = 28.87\,\Omega$ (Eqn. 2.16). All other elements shown in the Figures 2.28(a), 2.28(b) and 2.28(c) are ideal components.

The properties of the elements of the three different DTs are listed in table 2.5 and they show some interesting relations.

The optimum inductance of the single unit cells for the 3, 4 and 5 port as well as the optimum capacitances are all calculated by Eqn. 2.13 and Eqn. 2.14, respectively. In these equations, as well as for all other optimization equations presented in this thesis, the inductance is proportional and the capacitance inverse proportional to the load resistance. As described earlier, the resistive load of each unit cell is determined by the termination on the primary side (single ended side) divided by the number of ports on the secondary side. Therefore, twice the number of

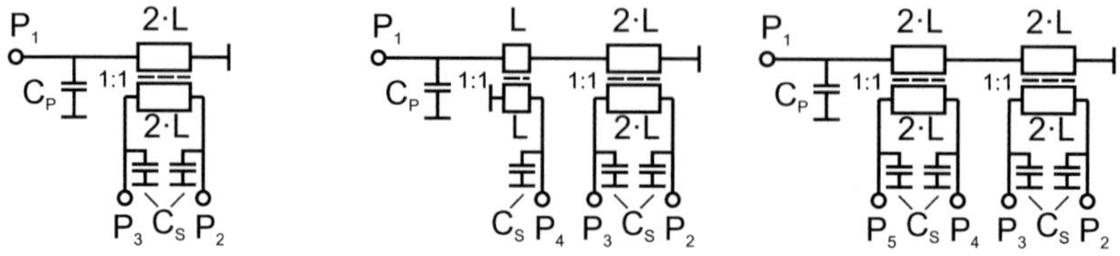

(a) 3 Port DT schematic ($50\,\Omega \leftrightarrow 2 \times 25\,\Omega$)

(b) 4 Port DT schematic ($50\,\Omega \leftrightarrow 3 \times 16.6\,\Omega$)

(c) 5 Port DT schematic ($50\,\Omega \leftrightarrow 4 \times 12.5\,\Omega$)

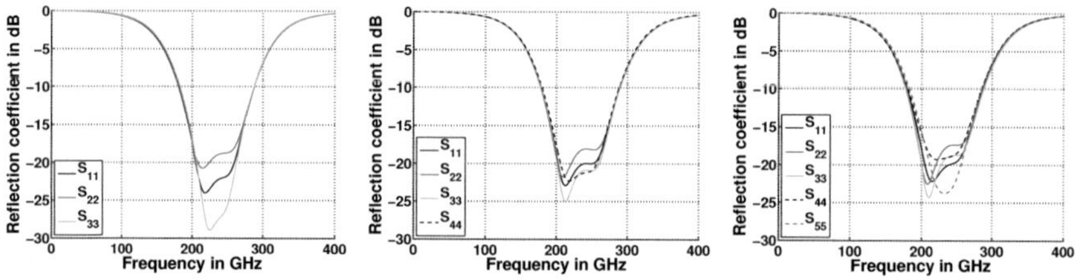

(d) 3 Port DT - Reflection coefficient

(e) 4 Port DT - Reflection coefficient

(f) 5 Port DT - Reflection coefficient

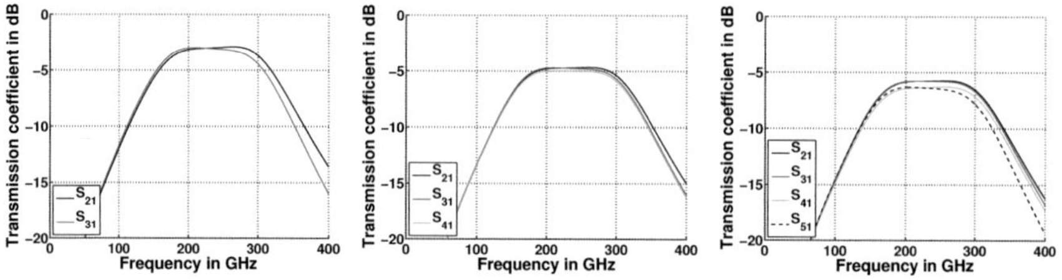

(g) 3 Port DT - Transmission coefficient

(h) 4 Port DT - Transmission coefficient

(i) 5 Port DT - Transmission coefficient

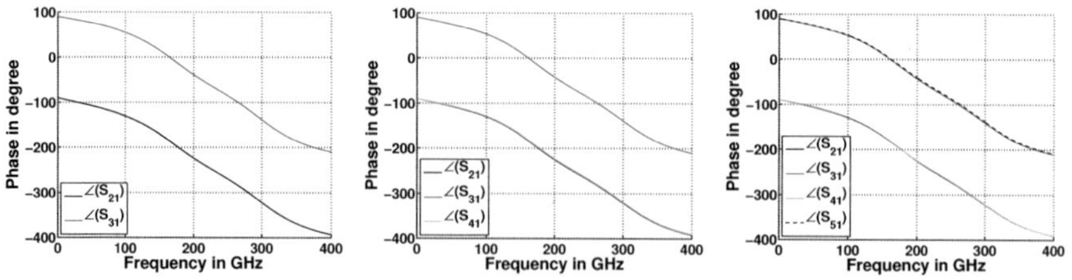

(j) 3 Port DT - Phase

(k) 4 Port DT - Phase

(l) 5 Port DT - Phase

Figure 2.28: Variation of the reflection and transmission coefficients of a 3-port, 4-port and 5-port DT. ($k_\mathrm{m} = 0.5$, $Q = \infty$ and $Z_{\mathrm{k}0} = 50\Omega$)

	L (pH)	$l_{unit\ cell}$ (%/λ)	l_{DT} (%/λ)	C_{s} (fF)	C_{p} (fF)
3 port (2 x unit cells)	10.6	4.3	8.6	47.7	23.85
4 port (3 x unit cells)	7.1	2.9	8.75	71.5	23.85
5 port (4 x unit cells)	5.3	2.2	8.8	95.4	23.85

Table 2.5: Parameters of a 3-, 4-, and 5-port DT based on coupled line sections with the same properties ($k_{\mathrm{m}} = 0.5$, $Z_{\mathrm{k0}} = 50\,\Omega$) but various lengths.

output ports leads to half the inductance (L_{p}) and twice the capacitance (C_{s}) per unit cell as visible in table 2.5 for the 3 and 5 port. The relation between the length of a shorted coupled line and its inductance was defined in Eqn. 2.18 and for small inductance values the length is proportional to the inductance ([BSMM13]: $\arctan(x) = x - \frac{1}{3}x^3 + \frac{1}{5}x^5 - ...| \, |x| < 1$). Hence, the overall length of a DT in principle does not depend on the number of unit cells, i.e. the number of output ports, since N times more ports also mean N times smaller port impedance and therefore about N times shorter coupled lines. That the overall length of the DT is independent of the number of unit cells, i.e. DT ports, is also the reason why the primary capacitance (C_{p}) is identical for all three transformers compared in Fig. 2.28.

By looking at the reflection and transmission coefficients as well as the phase of the 3, 4 and 5 port DTs, it can be seen that the overall performance does not change significantly as the number of ports is increased. The bandwidth in terms of RL and IL stays more or less the same for all three configurations and even though the IL starts to vary with an increasing number of unit cells/ports, the variations stay within an acceptable range for the coupled line applied in this example ($k_{\mathrm{m}} = 0.5$, $Z_{\mathrm{k0}} = 50\,\Omega$).

In the previous section it was shown that the efficiency as well as the bandwidth of a DT depend mainly on the coupling coefficient (k_{m}). The higher the coupling coefficient the better the performance in terms of efficiency and bandwidth of a single unit cell. However, as soon as there is not just one unit cell in the DT anymore but multiple, the performance of the DT can degrade in terms of phase and amplitude imbalance, if the coupling coefficient gets too high.

The optimum inductance in terms of efficiency or bandwidth depends largely on the coupling coefficient. As the coupling coefficient rises, so does the optimum inductance value. If the inductance goes up, the length of the coupled line, representing the inductance in a DT, has to increase as well. Therefore, the overall length of an optimized DT goes up, which, at a certain point, leads to a severe influence on the amplitude and phase balance between the single ports of the DT. To get an idea about the order of degradation that can be expected as k_m increases, Fig. 2.29(b) and Fig. 2.29(c) depict the maximum amplitude and phase difference between the ports of a 3 port all the way up to a 7 port DT optimized for maximally flat response at 240 GHz versus k_m. The characteristic impedance of the coupled line in this example is again set to $Z_{k0} = 50\,\Omega$.

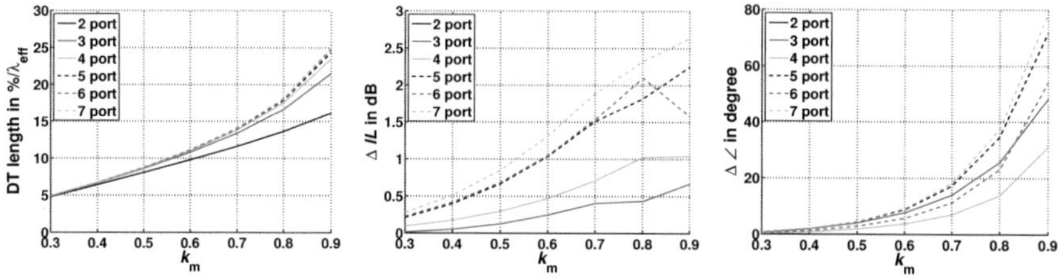

(a) DT length vs. k_m (b) DT IL difference vs. k_m (c) DT Phase difference vs. k_m

Figure 2.29: Influence of the coupling coefficient (k_m) on the phase and amplitude imbalance of the DT ports at 240 GHz. ($Z_{k0} = 50\Omega$)

In Fig. 2.29(a) the corresponding DT lengths versus k_m are plotted. Since the inductance of a coupled line is proportional to $\tan(l)$ the lengths vary slightly for high coupling coefficients, which lead to high inductance values. At a coupling coefficient of about 0.6 the maximum phase difference between the ports of the DTs becomes more than 10° and the maximum amplitude imbalance of the DTs with more than 5 ports is more than 1 dB.

This means that if phase and amplitude imbalance are not allowed to become worse than e.g. 10° or 1 dB, the maximum coupling factor or the number of ports is limited for a coupled line with a characteristic impedance scaling parameter of $Z_{k0} = 50\,\Omega$. However, if the characteristic impedance can be increased, it is possible to achieve a good amplitude and phase balance, even for high coupling coefficients.

Fig. 2.30 shows again the relative DT length as well as the maximum amplitude and phase imbalance, but this time versus Z_{k0}. The coupling coefficient (k_m) in all curves is 0.8.

As Z_{k0} increases, the lengths of the unit cells optimized for maximally flat response decrease, since the inductance and therefore the length of a coupled line is proportional to Z_{k0} (Eqn. 2.17). This leads to an improved amplitude and phase balance.

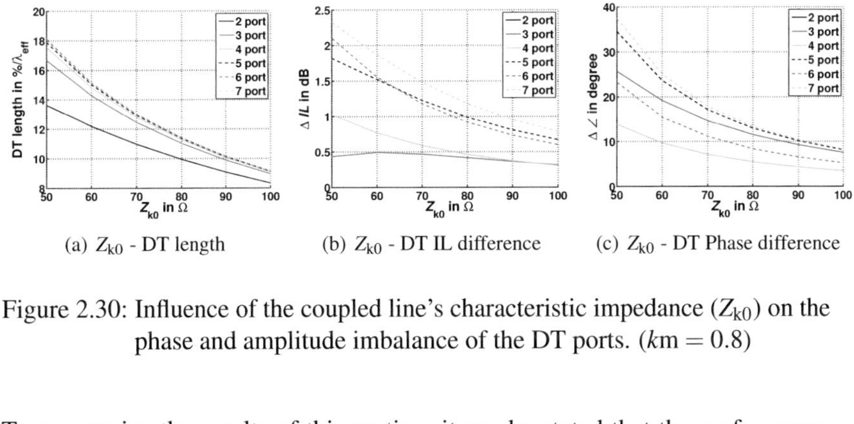

(a) Z_{k0} - DT length (b) Z_{k0} - DT IL difference (c) Z_{k0} - DT Phase difference

Figure 2.30: Influence of the coupled line's characteristic impedance (Z_{k0}) on the phase and amplitude imbalance of the DT ports. ($km = 0.8$)

To summarize the results of this section, it can be stated that the performance of a DT is more or less identical to the performance of a single DT unit cell, if the characteristic impedances of the applied coupled line in conjunction with its coupling coefficient result in an overall transformer length that is small compared to the guided wavelength. If this condition is fulfilled, the performance and also the size of a DT are more or less independent of the number of cascaded unit cells, i.e. DT ports.

2.7 Conclusion on chapter 2

In chapter 2, the optimization of the frequency response of a DT has been reduced to the optimization of a DT unit cell. The DT unit cell is represented by a simple ECM. Equations used in literature to optimize the efficiency of a DT/DAT have been introduced and used to estimate the maximum available efficiency of a DT. The evaluation of an efficiency optimized DT unit cell ECM over frequency revealed that the efficiency optimization equations always lead to a response, where the first resonance of the DT and the design frequency fall together. Since the DT unit cell and also the DT in general behave like a second order bandpass filter, this means that the efficiency optimization equations result in an unsymmetrical frequency response with regard to the optimization frequency.
Novel design equations, which facilitate the design of DTs with a maximum symmetrical BW, have been presented. These equations are derived from the lossless variant of the ECM, used to derive the equations for maximum efficiency. New design procedures and equations for the design of DTs with maximum symmetrical bandwidth as well as a maximally flat IL (Butterworth) or a RL level of better than 30 dB, 20 dB or 10 dB (Chebyshev) have been presented.
A comparison between the achievable symmetrical RL bandwidths of all four bandwidth optimization strategies was provided. It revealed that only for an aspired RL of more than approximately 20 dB, an optimization for exactly the desired minimum RL results in significant bandwidth improvement compared to an optimization for maximally flat IL for instance.
The influence of loss on the BW optimized DT responses was demonstrated. It was shown that while the loss degrades the efficiency at the design frequency more serious than for an optimization for maximum efficiency, it has almost no influence on the bandwidth of the DT. Its main effect is that it leads to a slope of the transmission coefficient within the passband of the DT.
To evaluate the eligibility of the optimization equations obtained from the lumped ECM for a distributed transformer, the lumped transformer in the original ECM was replaced by a pair of coupled lines, which led to an ECM that is closer to the real DT geometry.

While the new distributed ECM is too complex to derive optimization equations on its basis, it is a perfect candidate for their verification. The application of the design procedure for maximum symmetrical bandwidth proved to be successful for the coupled line based ECM. IL and RL showed the anticipated response and kept symmetrical around the design frequency.
It was also seen that the efficiency of the ECM based on a lossy coupled line instead of a lossy lumped transformer turned out to be much higher, even though the same quality factors for the primary and secondary transformer inductance were used. The reason for this was found to be the resistive coupling, which is neglected in the original lumped ECM.
Since the simple lumped ECM predicts a different efficiency than the distributed model, the efficiency optimization, based on the simple lumped ECM, fails for the distributed coupled line based model. This, on the other hand, means that the optimization of a real distributed transformer would also fail.
To figure out the reason for the deviation between both models, the ECM of a coupled line was investigated. By extracting the lumped model parameter of the coupled line, the severe resistive coupling was discovered.
Based on the ECM of a coupled line, which was extended to incorporate the resistive coupling, a new lumped ECM for the DT was developed. It served as origin for the development of a new efficiency equation as well as new optimization equations for DTs/DT unit cells with maximum efficiency.
By applying the new equations, the successful optimization of the distributed coupled line based ECM was demonstrated.
To get an idea about the efficiency values, which are possible for different coupling coefficients (k_{m}) and quality factors (Q) and to demonstrate the difference between the old and the new ECM based efficiency equations, a comparison of the maximum efficiency determined by the old and new model equations versus k_{m} and Q was presented. It was shown that due to the resistive coupling in DTs, higher efficiencies than those predicted by the original lumped ECM can be achieved. The higher the coupling coefficient and lower the quality factor, the higher the influence of the resistive coupling and therefore the difference between the old and new optimization equations.

At the beginning of the chapter the DT was reduced to a single unit cell. At its end the circle is closed by an investigation of the influence of the parameters of a single coupled line based DT unit cell on the phase and amplitude balance of the ports of a complete DT with N ports and therefore N DT unit cells. In general, the section proved that the optimization of a single DT unit cell results in an optimized DT build by multiple optimized DT unit cells. The DT shows the same frequency response as the single DT unit cells. The main condition that was found for a good amplitude and phase balance between the ports of a DT is that the size of the complete DT should be kept as small as possible. The imbalance increases with the size of the complete DT. The number of DT unit cells however has almost no influence on the overall length of the DT, since the length of a single DT unit cell scales with the number of ports, i.e. the load termination of the ports. Twice the number of DT unit cells also means twice the number of ports and therefore half the port impedance and DT unit cell length.
In the next chapter the findings of this chapter will be used to design DT based IMNs/OMNs.

3 Design and Implementation of Distributed Transformers

In the previous chapter the optimization of a DT as an integral part of an IMN or OMN was broken down to the optimization of a single DT unit cell. It was shown that the general performance of a DT is mainly determined by the behavior of the lumped or coupled line based DT unit cell. Up to now, there have been no restrictions regarding how the single elements of a DT are build. The unit cell and the DT could be constructed by any type of capacitors and mutually coupled inductors respectively any type of coupled lines. The parameters limiting the performance so far, are the coupling coefficient (k_{m}) and the quality factors (Q) of the inductors as well as the realizable characteristic impedances (i.e.: Z_{k0}) of the coupled line segments. Their influence on the response of a unit cell but also a DT as a cascadation of multiple unit cells, has been investigated in the previous chapter.

In this chapter a strategy for the design of a single DT unit cell and a complete DT will be presented. The necessary parameters for an optimization (i.e.: k_{m}, Q_{p}, Q_{s} and Q_{m}) will be extracted from S-parameters of EM simulations. These EM simulations implement coupled lines on an advanced Fraunhofer IAF 35 nm or 50 nm mHEMT process and abide their limitations. Various possible unit cell implementations and geometries will be compared for their eligibility as coupled line of a DT. The most promising unit cell implementation will be optimized for a maximally flat frequency response and maximum efficiency based on the equations of the previous chapter. Composed out of multiple of those unit cells, an entire DT will be designed and compared to EM simulations.

Finally an EM simulated 5 port DT will be compared to measurements.

3.1 Design strategy for a DT based IMN/OMN

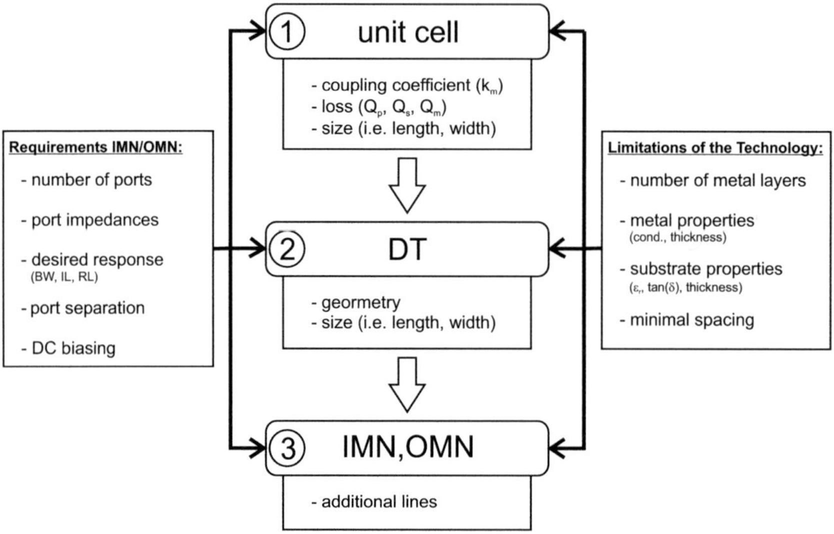

Figure 3.1: Design flow for the design of a DT based IMN/OMN.

As visualized in Fig. 3.1, the design of a DT based IMN/OMN mainly consists of the three phases:

- design of the DT unit cell
- design of the DT
- design of the IMN/OMN comprising the DT

All three design phases are influenced by the requirements for the IMN or OMN on one side and by the limitations and properties of the applied semiconductor process on the other.
During the design of the unit cell it has to be decided, whether the DT and therefore

the IMN/OMN shall be designed for maximal symmetrical bandwidth around the operation frequency or if an optimization for maximum efficiency returns a frequency response that has enough bandwidth for the desired application but obviously less loss. Depending on the number of parallel outputs and their desired port impedance, the number of unit cells and their length is determined. A lower port impedance results in a shorter coupled line as was already described in chapter 2. To maximize the efficiency and bandwidth, the coupling coefficient (k_m), characteristic impedances (Z_{k0}) as well as the quality factors (Q) of the coupled line have to be as high as possible. To achieve that, the geometry of the applied coupled lines has to be optimized for this purpose. Very often, this leads to the limits of the applied MMIC process in terms of minimum spacing and feature size in general. After the DT unit cells are optimized with respect to the requirements of the matching network as well as the limits of the technology, they have to be combined in a way that results in the right number of ports and also the right port impedances. This step is design phase two: the design of the DT.

The identical unit cells can be connected in various ways that all lead to more or less the same IL and RL response, but different geometries of the DT.

Various kinds of geometries and their properties regarding impedance transformation, signal distribution (port separation), port phase and so forth, will be discussed in a later section of this chapter.

The right decision regarding which metal layer of the applied process becomes the primary or secondary side of the transformer can ease the connection of the required parallel capacitors at each port of the transformer or the connection of transmission lines linking the transformer to its surrounding elements. The routing of lines as well as the design of the DT's capacitors is already part of the third phase: the design of the IMN/OMN comprising the DT.

The low impedance of the ports on one side of the transformer is usually an advantage, as it reduces the required matching effort of connecting the active elements (transistors), which show a very low input and output impedance at mmW frequencies. In a standard DT configuration however the different ports on the low impedance side of the transformer usually lie quite close together, which can be an issue, if the DT's parallel capacitors have to be quite big or if the circuits that shall be connected with the DT are spacious and therefore cannot be connected directly. Hence, it can be necessary to use low impedance lines to connect the low impedance ports to their destinations and also to optimize the geometry of the DT to support the spatial separation of the DT ports.Therefore, methods and geometries that enable a good spacial separation between the DT ports will be introduced in the sections ahead.

3.2 The Fraunhofer IAF 35 nm and 50 nm mHEMT process and its possible coupled line geometries

The first step of the design of a DT is the investigation of the possible coupled line structures, which build the core of each transformer. Fig. 3.1 lists a number of technology limitations that can have a decisive influence on the possible DT unit cell performance. All these parameters are defined by the applied MMIC technology.

Within this thesis all DTs will be built on Fraunhofer IAF's highly advanced 35 nm and 50 nm mHEMT process [LTK^+10, LTM^+12, LTD^+13] technology.

The layer stack of the IAF processes is depicted in Fig. 3.2, where all possible stack configurations can be seen. As visible in the figure, the technology provides the possibility of having two metal layers (Metal 1 (Met1) and Galvanic Metal (MetG)) on top of each other, which are either connected directly (Fig. 3.2 - c) or isolated by a thin Silicon Nitride (SiN) dielectric layer (Fig. 3.2 - e). It is also possible to route the top metal layer (MetG) over the metal layer (Met1) below in form of air bridges (Fig. 3.2 - d).

The minimum feature size and minimum spacing between the metals is listed in table 3.1 and the properties of the dielectrics are given in table. 3.2.

To convey an idea of how the various possible layer stacks look like in reality, Fig. 3.3 shows a Scanning Electron Microscope (SEM) picture of some of the passive elements available in the applied MMIC technology.

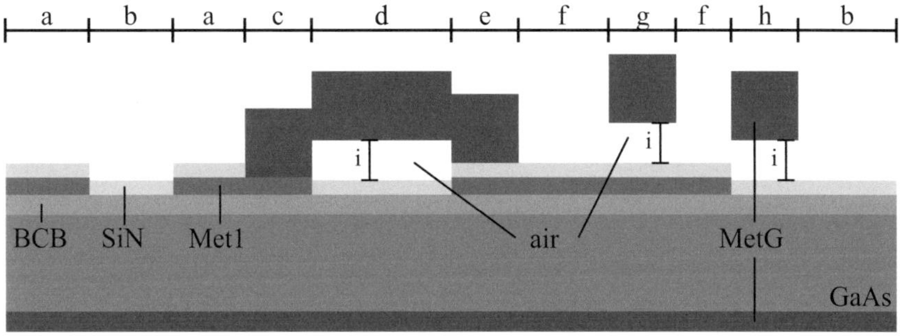

Figure 3.2: Possible combinations of the available metal and dielectric layers. [Rei13]

Table 3.1: Process limitations regarding spacing and metal layer size.

parameter	description	range
a	min. Met1 width	$a \geq 2\,\mu m$
b	min. Met1 separation	$b \geq 2\,\mu m$
c	min. MetG width	$c \geq 4\,\mu m$
d	air bridge length	$4\,\mu m \leq d \leq 100\,\mu m$
g	air bridge width	$4\,\mu m \leq g \leq 20\,\mu m$
f	min. MetG separation	$f \geq 4\,\mu m$
i	air bridge height	$g = 1.6\,\mu m$

Table 3.2: Process layer properties.

layer	**thickness**	**permittivity ε_r/ conductivity σ**	$\tan\delta$
GaAs	50 µm	$\varepsilon_r = 12.94$	$\tan\delta = 0.006$
SiN	0.25 µm	$\varepsilon_r = 7.5$	-
BCB	0.35 µm	$\varepsilon_r = 2.65$	$\tan\delta = 0.002$
Met1	0.3 µm	$\sigma = 3.3 \cdot 10^7 S/m$	-
MetG	2.7 µm	$\sigma = 3.7 \cdot 10^7 S/m$	-

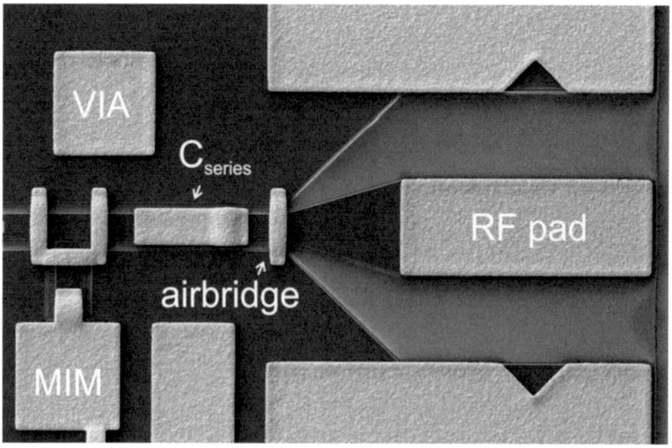

Figure 3.3: SEM photo of the IAF MMIC passives [TLM+08].

3.3 Possible unit cell configurations and their eligibility for DTs

As mentioned before, the first step towards a DT is the design of the unit cell's coupled line transformer. The objective is to find the best coupled line geometry in terms of coupling coefficient (k_{m}), loss (Q, i.e. α) and characteristic impedance ($\mathrm{Z_e}$, $\mathrm{Z_o}$, i.e. $\mathrm{Z_c}$, Z_π). The most common configurations, which are supported by the applied IAF process are depicted in Fig. 3.4(a) - 3.4(c). Here, Fig. 3.4(a) and Fig. 3.4(b) show two edge coupled configurations. The first uses just Met1, which enables a minimum separation (sep) between the two coupled lines of 2 µm, the other one uses lines, which consist of Met1 and MetG, which, due to process design rules, results in a wider gap of at least 4 µm. Unlike the first two configurations the third configuration is not edge but broadside coupled. The coupled line pair consists of one line implemented in Met1 and the other one implemented in MetG. Both are stacked on top of each other and separated by a 1.85 µm (0.25 µm SiN + 1.6 µm air) gap. Note that while configuration one and two have a symmetry plane right in the middle between both lines, configuration three does not have such a symmetry. Therefore, configuration one and two are also called symmetrical coupled lines and configuration three is called asymmetrical coupled line. The theory of symmetrical coupled lines is a subset of the equations for the asymmetric coupled lines.

To find out which of the three configurations is best suited as core of a DT unit cell, short coupled lines of each configuration have been simulated in an EM field solver [Wei03]. As described in section 2.5, the S-parameters obtained from these simulations can be converted into per unit-length quantities ($\mathrm{R}', \mathrm{L}', \mathrm{G}', \mathrm{C}'$). These quantities are plotted for all three configurations vs. the ground separation (*gnd_sep*) and the line width (*cwidth*) in appendix F. Based on these parameters the relevant properties (magnetic coupling coefficient (k_{m}), loss (i.e. α) and the characteristic impedances (e.g.: $\mathrm{Z_e}$, $\mathrm{Z_o}$, $\mathrm{Z_c}$ and Z_π) of the three coupled line configurations can be calculated. The necessary equations that describe the relation between all types of 4 port parameters (Z-, Y-, S-, ABCD-, etc.) of an asymmetrically coupled line and its per unit-length quantities can be found in [Tri75, MBBH07] and for the sake of completeness are repeated in appendix E.

A coupled line, as it is considered in this thesis, consists of two lines with independent potentials referred to another reference potential (i.e. ground (GND)). Having three independent potentials means that the coupled line supports two propagating modes. For the symmetrical coupled line these modes are called *even* and *odd* modes and for the asymmetrical coupled line one refers to the c and π modes, respectively. Both modes can have different propagation constants (γ), which means that the loss (α) as well as the propagation velocity of both modes can be different (e.g. in quasi TEM lines). To evaluate the loss of the three coupled line configurations, the mean value of the real part (α) of the complex propagation constants (γ) of both coupled line modes is a good indicator and therefore will be used to select the best coupled line geometry.

Besides the loss of the coupled line, which is directly related to the quality factors of the DT unit cell, the magnetic coupling coefficient is another important coupled line parameter. The higher the magnetic coupling the higher the transformer unit cell's efficiency and bandwidth. In chapter 2, k_{m} was always set to a fixed value to compare various optimization methods under the same conditions. Now that real coupled line geometries are compared, their coupling coefficients have to be calculated because they depend on the geometry. This calculation can be done on the basis of per unit-length quantities. For this, the ratio between the mutual inductance (M') and the self inductance (L'_1, L'_2) of the two coupled lines is calculated:

$$k_{\mathrm{m}} = \frac{M'}{\sqrt{L'_1 L'_2}} \tag{3.1}$$

Alternatively, the coupling coefficient can be determined on the basis of two port Z-parameters of a short line segment ($l << \lambda/4$) shorted to GND:

$$k_{\mathrm{m}} = \sqrt{\frac{\mathrm{Im}\{Z_{12}\}\,\mathrm{Im}\{Z_{21}\}}{\mathrm{Im}\{Z_{11}\}\,\mathrm{Im}\{Z_{22}\}}} \tag{3.2}$$

A shorted coupled line segment can be obtained from a coupled line 4 port by shorting one end of the coupled line.

The last important parameter to determine, which of the proposed coupled line geometries is best for an implementation as DT unit cell, is the characteristic impedance.

In chapter 2 the response of a DT unit cell was optimized by calculating the optimum inductance and capacitance for a given set of coupling coefficient, load termination and quality factors. In the last section of chapter 2 it was demonstrated that if the inductors are implemented as coupled lines, the required port inductance can lead to coupled lines that are not small compared to the guided wavelength anymore and therefore lead to an unwanted amplitude and phase imbalance. To mitigate this issue, the characteristic impedances of the modes of the coupled line have to be high enough, since the mean value of the characteristic impedance of the two propagating modes of a symmetrical line is directly proportional to the port inductance, as already described in Eqn. 2.17. Therefore, a high mean value leads to a short coupled line, which reduces any deviation from the optimal lumped DT behavior. A symmetrical coupled line has two characteristic impedances (Z_e, Z_o), which determine the ratio between the voltage and current waves on each of the coupled lines for the two modes. Since both lines are completely symmetrical with respect to their geometry, the ratios between voltage and current waves on both lines for both modes are identical (Eqn. 3.3).

$$\begin{aligned} Z_e = \frac{U_{1,e}^+}{I_{1,e}^+} = \frac{U_{1,e}^-}{I_{1,e}^-} &= \frac{U_{2,e}^+}{I_{2,e}^+} = \frac{U_{2,e}^-}{I_{2,e}^-} \\ Z_o = \frac{U_{1,o}^+}{I_{1,o}^+} = \frac{U_{1,o}^-}{I_{1,o}^-} &= \frac{U_{2,o}^+}{I_{2,o}^+} = \frac{U_{2,o}^-}{I_{2,o}^-} \end{aligned} \tag{3.3}$$

If the coupled line is asymmetrical, this is not true anymore, which means that for an asymmetrically coupled line there are actually four characteristic impedances (Z_{c1}, $Z_{\pi 1}$, Z_{c2} and $Z_{\pi 2}$) describing the ratios between the voltage and current waves on both lines (Eqn. 3.4).

$$
\begin{aligned}
Z_{\mathrm{c},1} = \frac{U^{+}_{1\mathrm{c}}}{I^{+}_{1,\mathrm{c}}} = \frac{U^{-}_{1,\mathrm{c}}}{I^{-}_{1,\mathrm{c}}} \quad &\neq \quad \frac{U^{+}_{2\mathrm{c}}}{I^{+}_{2,\mathrm{c}}} = \frac{U^{-}_{2,\mathrm{c}}}{I^{-}_{2,\mathrm{c}}} = Z_{\mathrm{c},2} \\
Z_{\pi,1} = \frac{U^{+}_{1,\pi}}{I^{+}_{1,\pi}} = \frac{U^{-}_{1,\pi}}{I^{-}_{1,\pi}} \quad &\neq \quad \frac{U^{+}_{2,\pi}}{I^{+}_{2,\pi}} = \frac{U^{-}_{2,\pi}}{I^{-}_{2,\pi}} = Z_{\pi,2}
\end{aligned}
\tag{3.4}
$$

Since the characteristic impedances of the two modes for both lines are not identical, the amplitudes of the voltage and current waves on both lines are not identical either. Their ratio for the two modes can be defined as follows:

$$
\begin{aligned}
R_{\mathrm{c}} &= \frac{V^{+}_{2,\mathrm{c}}}{V^{+}_{1,\mathrm{c}}} = \frac{V^{-}_{2,\mathrm{c}}}{V^{-}_{1,\mathrm{c}}} = -\frac{I^{+}_{1,\pi}}{V^{+}_{2,\pi}} = -\frac{I^{-}_{1,\pi}}{I^{-}_{2,\pi}} \\
R_{\pi} &= \frac{V^{+}_{2,\pi}}{V^{+}_{1,\pi}} = \frac{V^{-}_{2,\pi}}{V^{-}_{1,\pi}} = -\frac{I^{+}_{1,\mathrm{c}}}{V^{+}_{2,\mathrm{c}}} = -\frac{I^{-}_{1,\mathrm{c}}}{I^{-}_{2,\mathrm{c}}}
\end{aligned}
$$

To figure out how the four characteristic impedances of an asymmetrically coupled line are related to the port inductance of the two lines, the four port parameters of an asymmetrical coupled line can be used. In appendix E the Y-parameters of an asymmetrically coupled line are given in Eqn. E.17. By picking the correct set of parameters from the 4 port, the Y-parameters of a two port DT unit cell (one side of the coupled line is short circuited to GND) can be extracted. If the Y-parameters are converted to Z-parameters, Eqn. 2.19 can be applied to calculate the two port inductances of a DT unit cell, which turn out to be as follows:

$$
\begin{aligned}
L_1 &= \frac{\mathrm{Im}\, Z_{11}}{\omega} = \frac{1}{1-\frac{R_{\mathrm{c}}}{R_{\pi}}} Z_{\mathrm{c},1} \tanh \gamma_{\mathrm{c}} l + \frac{1}{1-\frac{R_{\pi}}{R_{\mathrm{c}}}} Z_{\pi,1} \tanh \gamma_{\pi} l \\
L_2 &= \frac{\mathrm{Im}\, Z_{22}}{\omega} = \frac{1}{1-\frac{R_{\pi}}{R_{\mathrm{c}}}} Z_{\mathrm{c},2} \tanh \gamma_{\mathrm{c}} l + \frac{1}{1-\frac{R_{\mathrm{c}}}{R_{\pi}}} Z_{\pi,2} \tanh \gamma_{\pi} l
\end{aligned}
\tag{3.5}
$$

Note that Eqn. 3.5 becomes identical to Eqn. 2.17, which defined the port inductance for a symmetrical coupled line, if $R_c = 1$ and $R_\pi = -1$. In this case the c mode of the asymmetrical coupled line turns into the *even* mode of the symmetrical coupled line and the π mode becomes the *odd* mode.

As it turns out during the analysis of the various geometries of the asymmetrical coupled line depicted in Fig. 3.4(c), the difference between the inductance of the two ports of a DT is always very small, which is why the mean value of both port inductances will be used within any optimization of an asymmetrical coupled line DT unit cell in this thesis.

To compare the influence of the characteristic impedance of the symmetrical coupled lines with that of the asymmetric one, the mean value of the weighted characteristic impedance of both lines of the asymmetrical coupled line ($\overline{\frac{Z_c+Z_\pi}{2}}$) will be evaluated. It is defined in Eqn. 3.6 and it equals the mean of the characteristic impedance of the even and odd mode of a symmetrical coupled line, if $Z_{c,1} = Z_{c,2} = Z_e$, $Z_{\pi,1} = Z_{\pi,2} = Z_o$ and $R_c = 1$ and $R_\pi = -1$.

$$\overline{\frac{Z_c + Z_\pi}{2}} = \frac{\left(\frac{1}{1-\frac{R_c}{R_\pi}} Z_{c,1} + \frac{1}{1-\frac{R_\pi}{R_c}} Z_{\pi,1}\right) + \left(\frac{1}{1-\frac{R_\pi}{R_c}} Z_{c,2} + \frac{1}{1-\frac{R_c}{R_\pi}} Z_{\pi,2}\right)}{2} \tag{3.6}$$

Now that the figures of interest for the assessment of the three different coupled line geometries are defined, the next step is the evaluation of the results. In Fig. 3.4 all three different parameters, the mean of the attenuation coefficients of both propagating modes, the coupling coefficient and the mean of the characteristic impedance, for all three different coupled lines are plotted versus various line widths (*cwidth*) and ground separations (*gnd_sep*). The separation between the lines is always set to its minimum value supported by the applied IAF process. This means that the separation for configuration 1 (only Met1 lines) is 2 µm and for configuration 2 (lines consist of Met1 and MetG) 4 µm. In configuration 3 the distance between the lines is 1.85 µm (0.25 µm SiN + 1.6 µm air - (Fig. 3.2 - g)) and both lines are exactly on top of each other.

As mentioned before, the objective is to find the line with the least mean loss (smallest α), highest mean characteristic impedance (Z) and highest mean magnetic coupling coefficient (k_m).

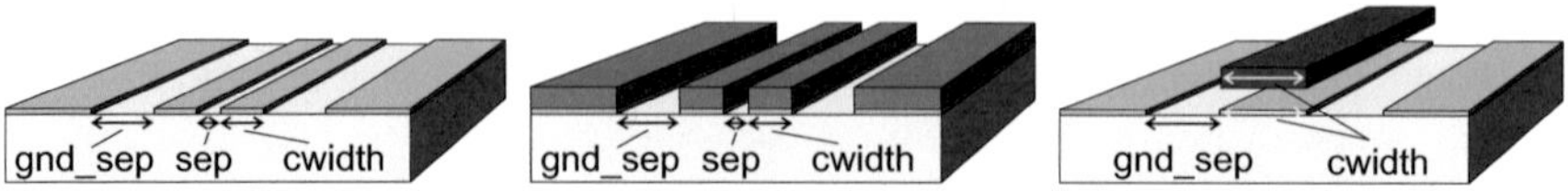

(a) Symmetrically coupled line (b) Symmetrically coupled line (c) Asymmetrically coupled line

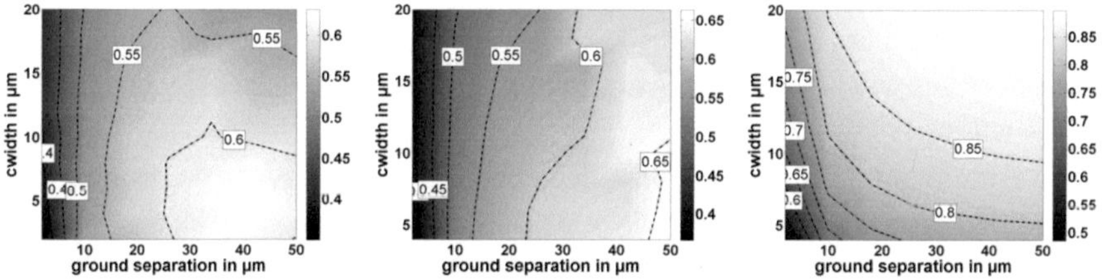

(d) k_{m} (e) k_{m} (f) k_{m}

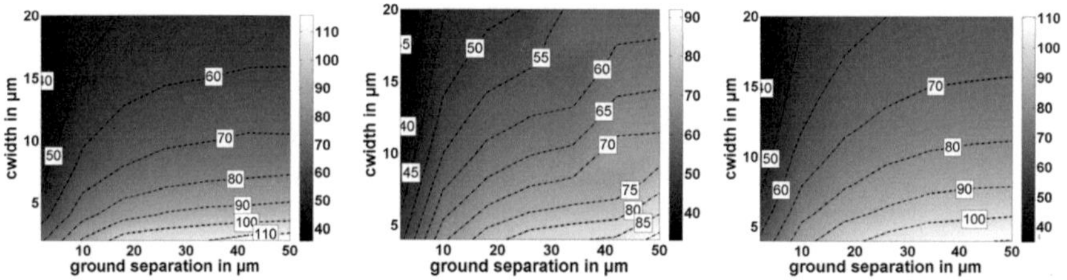

(g) $\frac{Z_{\mathrm{e}}+Z_{\mathrm{o}}}{2}$ in Ω (h) $\frac{Z_{\mathrm{e}}+Z_{\mathrm{o}}}{2}$ in Ω (i) $\frac{Z_{\mathrm{c}}+Z_{\pi}}{2}$ in Ω

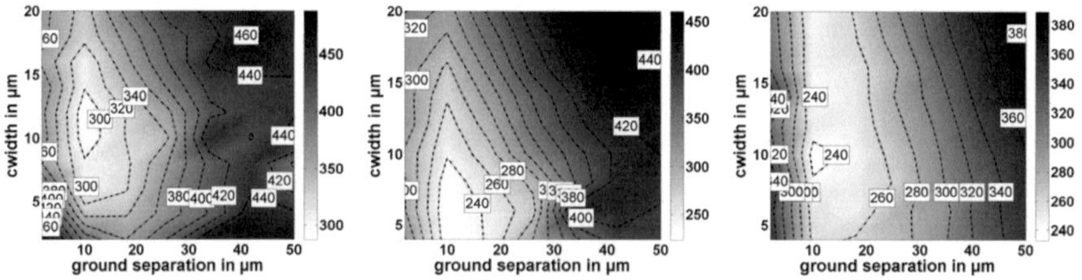

(j) $\frac{\alpha_{\mathrm{e}}+\alpha_{\mathrm{o}}}{2}$ in $\frac{1}{\mathrm{m}}$ (k) $\frac{\alpha_{\mathrm{e}}+\alpha_{\mathrm{o}}}{2}$ in $\frac{1}{\mathrm{m}}$ (l) $\frac{\alpha_{\mathrm{c}}+\alpha_{\pi}}{2}$ in $\frac{1}{\mathrm{m}}$

Figure 3.4: Comparison between the important line parameters of two symmetrical and one asymmetrical coupled line configuration at 240 GHz.

The second row of sub figures (Fig. 3.4(d) - 3.4(f)) shows k_m vs. *cwidth* and *gnd_sep* for all three coupled line geometries. According to Eqn. 3.1 the coupling coefficient increases as the ratio between the mutual inductance (L_m) and the self inductance (L_1, L_2) of the two lines increases. Therefore, the lines should be as close as possible (high L_m) and the ground metal should be as far away as possible (low L_1, L_2). Obviously, the magnetic coupling coefficient of all three geometries increases as the ground separation increases, which means that the ratio between the mutual inductance and the self inductance of the two lines increases (Eqn. 3.1) as expected. The influence of the line width (*cwidth*) on the coupling coefficient is different for the symmetrical (config. a, b) and the asymmetrical coupled lines (config. c). For ground separations, which are decisively larger than the separation between both lines, the coupling coefficient of the symmetrical lines decreases with an increasing line width and the coupling coefficient of the asymmetrical coupled line increases instead. This means that as *cwidth* increases so does the ratio between the mutual and self inductance of the asymmetrical line, while it decreases for the symmetrical lines. Note that in general k_m of the asymmetrically coupled line exceeds the value of the symmetrical one significantly. Hence, from a coupling coefficient point of view, the asymmetrical coupled line is superior.

The mean of the characteristic impedances for all three configurations is shown in Fig. 3.4(g) - 3.4(i). Recall that the objective is a high characteristic impedance as it leads to a short DT unit cell during optimization, which means less amplitude and phase imbalance for a DT composed out of these unit cells. As visible, the characteristic impedance of all three coupled line configurations increases with an increase of the ground separation. In terms of *cwidth*, narrower lines result in a higher characteristic impedance. The absolute values of configuration a and c are more or less in the same range. Compared to them, configuration b has a lower characteristic impedance. Therefore, if it was for the characteristic impedance, line a or c would be chosen.

In the last row of Fig. 3.4 the mean of the attenuation constant α for both modes of all three coupled line configurations are plotted versus the geometry parameters. All three coupled lines show a minimum in their attenuation at a ground separation of about 10 µm and depending on the configuration at various *cwidth* values. Configuration a shows the highest attenuation coefficients, which does not surprise, since it only uses the thin (300 nm thick) Met1 layer for all of its line and ground structures. Configuration b and c have more or less the same absolute attenuation

value. In general the attenuation coefficients of all three coupled lines are very similar. If one has to choose, configuration b and c are slightly superior compared to configuration a.
After considering all three configurations in terms of coupling coefficient, characteristic impedance and attenuation coefficient, the asymmetrical coupled line turns out to be the best choice for the applied process. It has by far the highest coupling coefficient and a characteristic impedance and loss that are close to the highest and lowest of all three configurations, respectively.
Thus, based on the asymmetrical coupled line configuration, the optimization equations of chapter 2 will be verified by designs of multiple DT based matching networks presented later in this chapter.

3.4 Possible DT geometries

Now that the most promising unit cell geometry has been chosen, the next step is the design of the DT by cascading the unit cells in a way that fulfills the requirements for the matching network (Fig. 3.1) and abides by the design rules of the applied MMIC process. For clarity reasons, the DTs described in this section all operate as part of an IMN, which has one port on the primary and N ports on the secondary side. All the conclusions obtained under this assumption are also valid for a DT as part of an OMN, where the primary side would have N simultaneously excited ports, while the secondary side is single-ended.
In principle the unit cells can be connected in whatever geometry suits best the design of the IMN. The only limitation is that the unit cell geometry should not be altered significantly by its arrangement in the DT, as this would change its behavior and therefore lead to a frequency response of the DT that deviates from the optimized frequency response of the single DT unit cell.
The major IMN properties, that have to be considered for the design of the DT are the desired impedance at the in- and output ports of the network, the number of ports on either side of the DT, the required phase relation between those ports and the desired spacial separation from each other.
A single DT unit cell, as it is defined in this thesis, performs no impedance transformation. It consists of the coupled line unit cell core (1:1 transformer),

whose four ports are either short circuited to gnd, connected to another unit cell or act as ports of the DT. In general two or even three of the four ports are either grounded or connected to another unit cell, while the remaining one or two ports become ports of the DT.
To the first extent, the DT unit cell is symmetrical, which means that there are two possible ways to connect the ports of a DT and still end up with a magnetic transformer. Both of these configurations are depicted in table 3.3-I,II. If the unit cell is assumed to be ideal (i.e. loss and phase variation due to the length of the coupled line are neglected and $k_m = 1$), the impedance of both ports in both DTs have to be identical and the magnitude of the transmission coefficient ($S_{21,12}$) is one. However, depending on which ports of the coupled line unit cell are chosen as ports of the DT, the phase of the transmitted signal is either 0° or 180° as indicated in the table.

Table 3.3: Possible unit cell connections.

	geometry	port impedance	transmission coefficients
I	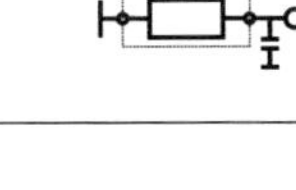	$Z_{P1} = Z_{P2}$	$S_{21} = S_{12} = +1$
II	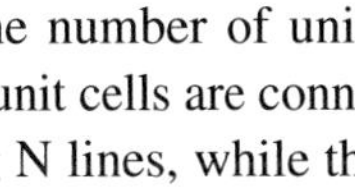	$Z_{P1} = Z_{P2}$	$S_{21} = S_{12} = -1$

By combining multiple DT unit cells, the DT can transform the impedance depending on the number of unit cells (N) and their way of combination. For instance, if N unit cells are connected in parallel the current at the input of the DT is split among N lines, while the voltage on all lines remains identical (current combiner/divider). Therefore, the impedance on the secondary side is increased by N (parallel connection - table. 3.4-V) because each unit cell (1:1 transformer) sees N times the input impedance. On the other hand, if the current in all unit cells is the same but the input voltage drops not just over one but multiple (N) unit cells (voltage combiner/divider), the impedance is decreased (series connection - table. 3.4-III, IV). In this case each unit cell sees only $1/N$ of the input impedance.

By using not only unit cells in parallel or in series but a combination of both, various transformation ratios can be achieved. For example an identical amount of parallel and series combinations leads to a DT with no impedance transformation at all (table 3.4-VI).

Table 3.4: Examples of possible unit cell combinations I.

† *Port 2 to port 5 have to be excited simultaneously.*

	geometry	port impedance	transmission coefficients
III	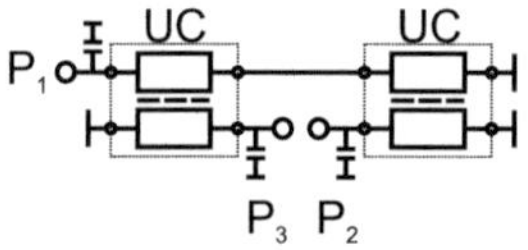	$\frac{1}{2} \cdot Z_{P1} = Z_{P2}{}^{\dagger} = Z_{P3}{}^{\dagger}$	$S_{21} = S_{12} = +\frac{1}{\sqrt{2}}$ $S_{31} = S_{13} = -\frac{1}{\sqrt{2}}$
IV	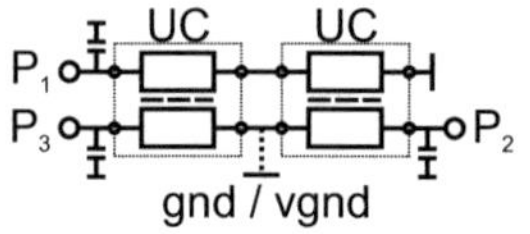	$\frac{1}{2} \cdot Z_{P1} = Z_{P2}{}^{\dagger} = Z_{P3}{}^{\dagger}$	$S_{21} = S_{12} = -\frac{1}{\sqrt{2}}$ $S_{31} = S_{13} = +\frac{1}{\sqrt{2}}$
V	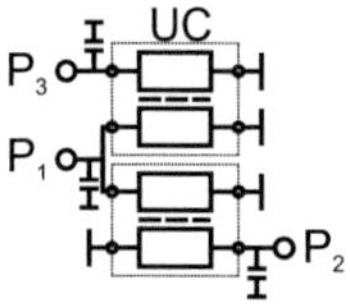	$2 \cdot Z_{P1} = Z_{P2}{}^{\dagger} = Z_{P3}{}^{\dagger}$	$S_{21} = S_{12} = -\frac{1}{\sqrt{2}}$ $S_{31} = S_{13} = +\frac{1}{\sqrt{2}}$
VI	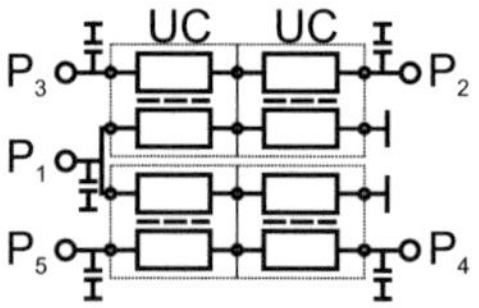	$Z_{P1} = Z_{P2}{}^{\dagger} = Z_{P3}{}^{\dagger}$ $= Z_{P4}{}^{\dagger} = Z_{P5}{}^{\dagger}$	$S_{21} = S_{12} = -\frac{1}{2}$ $S_{31} = S_{13} = +\frac{1}{2}$ $S_{41} = S_{14} = -\frac{1}{2}$ $S_{51} = S_{15} = +\frac{1}{2}$

Eqn. 3.7 gives the impedance transformation ratio between the single-ended input port on one and the multiple output ports on the other side of the designed DT.

$$\frac{Z_{\mathrm{IN}}}{Z_{\mathrm{OUT}}} = \tau = \frac{\text{number of serial unit cells}}{\text{number of parallel unit cells}} \tag{3.7}$$

The design decision, how many unit cells have to be used depends on the number of required parallel ports, their impedance and the necessary phase relation between them. The various properties of a DT can be tailored, by combining multiple unit cells in series or in parallel. Depending on which ports of the unit cell elements are short circuited or implemented as DT ports, the ports of the DT can have 0° or 180° phase shift between each other. As an example table 3.5-VII shows a seven port DT.

Operated as an IMN, the DT's input signal provided at port 1 is split into six output signals at the ports two to seven. While the magnitude of all six output ports is identical, the phase of the output signals is either 0° or 180° depending on the orientation of the DT unit cells. The DT has two parallel branches with each of them consisting of three unit cells in series. By evaluating Eqn. 3.7 this leads to an impedance transformation ratio of $\frac{Z_{\mathrm{P1}}}{Z_{\mathrm{P2-P7}}} = \frac{3}{2}$. Note that, due to the fact that the two parallel branches of the DT are fed with identical signals (in phase), the phase alternation between consecutive ports is interrupted between port four and five. To get a continuous alternation, the orientation of the unit cells of one branch can be adjusted as proposed in table 3.5-VIII. Theoretically, any sequence of port phases is possible. Due to the fact that some of these sequences require real and some only virtual gnd connections, some sequences are better suited for a real implementation on an MMIC process than others.

The last DT design consideration, which shall be mentioned, even though it is not pursued any further in this thesis, is the possibility to connect multiple high impedance ports to a single low impedance DT port. This way, the number of parallel ports is multiplied at the expense of higher impedances at these ports. Even though it seems obvious, it can be advantageous to keep this in mind for designs, where the alternating phase between neighboring ports as well as low impedances are not required, but the DT shall be employed as a compact, broadband, multiport balun. A balun consisting of a DT with one single ended port on one and four differential ports on the other side is depicted in table 3.5-IX.

Table 3.5: Examples of possible unit cell combinations II.
† *Port 2 to port 7 have to be excited simultaneously.*

	geometry	**port impedance**	**transmission coefficients**
VII	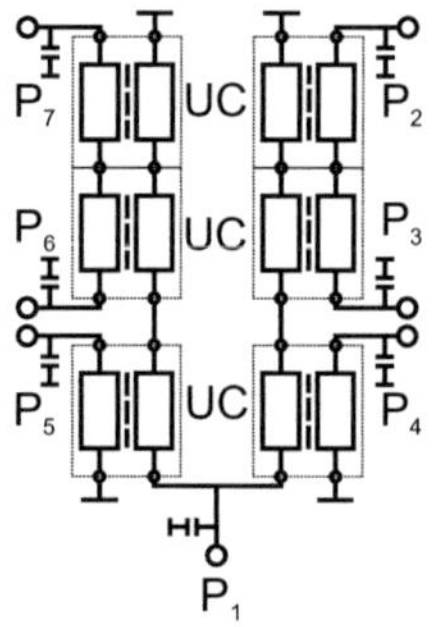	$\frac{2}{3} \cdot Z_{P1} = Z_{P2-P7}$ †	$S_{21} = S_{12} = -\frac{1}{\sqrt{6}}$ $S_{31} = S_{13} = +\frac{1}{\sqrt{6}}$ $S_{41} = S_{14} = -\frac{1}{\sqrt{6}}$ $S_{51} = S_{15} = -\frac{1}{\sqrt{6}}$ $S_{61} = S_{16} = +\frac{1}{\sqrt{6}}$ $S_{71} = S_{17} = -\frac{1}{\sqrt{6}}$
VIII	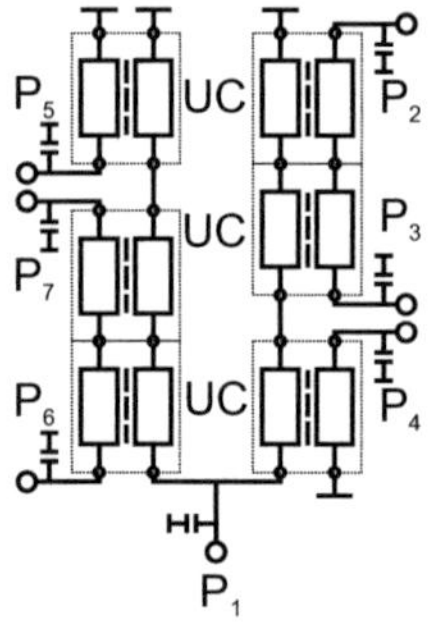	$\frac{2}{3} \cdot Z_{P1} = Z_{P2-P7}$ †	$S_{21} = S_{12} = -\frac{1}{\sqrt{6}}$ $S_{31} = S_{13} = +\frac{1}{\sqrt{6}}$ $S_{41} = S_{14} = -\frac{1}{\sqrt{6}}$ $S_{51} = S_{15} = +\frac{1}{\sqrt{6}}$ $S_{61} = S_{16} = -\frac{1}{\sqrt{6}}$ $S_{71} = S_{17} = +\frac{1}{\sqrt{6}}$
IX	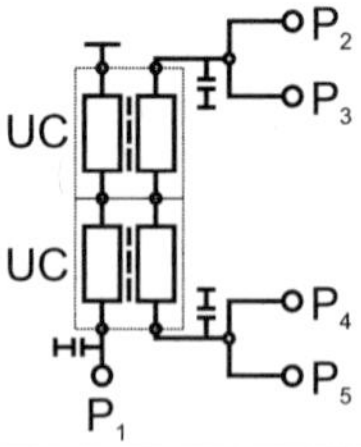	$Z_{P1} = Z_{P2-P4}$ †	$S_{21} = S_{12} = -\frac{1}{2}$ $S_{31} = S_{13} = -\frac{1}{2}$ $S_{41} = S_{14} = +\frac{1}{2}$ $S_{51} = S_{15} = +\frac{1}{2}$

For DTs as part of a PA matching network the voltage combiner with N unit cells in series on the single-ended side and in parallel on the differential side is the prevailing DT configuration. Unlike the current combiner type, it combines the property of a power combiner/divider with an impedance transformation from low impedances on the multiport side towards high impedances on the single-ended side and vice versa. It therefore can be applied beneficially in PAs with low impedance gain cells (e.g. transistors) and hence will be further investigated in this thesis.
Now that the effects of cascading multiple DT unit cells in various ways have been described, a couple of thoughts on the positioning of the unit cells respectively the overall geometry of the DT shall be given next.
In the introduction it was described that the first designed DATs all used circular geometries [AKRH02b]. The advantage of the circular DAT geometry is that in a circle always two active elements with opposite phase (180° phase difference) are right next to each other, which enables the use of push-pull transistor pairs and therefore eases the issue of lossy ground connections on Si based processes (virtual grounds). A problem of these circular geometries however, was the unwanted coupling of the circular transformer to its surrounding elements like the IMN, but also the other side of the transformer's circular geometry. Hence, it was necessary to investigate, i.e. simulate, the DAT as a whole. By adding a defined ground shield in close vicinity to the DAT's inductors to restrict the magnetic field by allowing a defined current return path [PG07], this issue could be reduced in later circular DATs. The restriction of the sphere of action of the DAT made it possible to optimize and evaluate single transformer unit cells, which is also why all the investigations and considerations presented in this thesis by now are valid. If the DT unit cells are connected together to design the entire DT, the geometry of the single unit cell should be altered as little as possible to maintain its response. However, depending on the ratio between the length of a single optimized unit cell and its width, this can become difficult for a circular geometry.
The unit cell that will be used for the DT design in this thesis is the previously selected asymmetrical coupled line (configuration c) depicted in Fig. 3.4(c). Its width is mainly determined by the ground separation, which is the spacing between the coupled line pair in the middle and ground (Met1) to its left and right. In the mmW frequency range, if a wide ground separation is chosen, the length of a

single unit cell will be in the same range as its ground separation or only slightly longer. This means that it is impossible to use a wide ground separation in a circular geometry as it could only be realized by severe changes to the unit cell's environment, leading to a deviation of the response of the unit cell. If the length of the single unit cell is not long compared to its width, the part of the unit cell where its (quasi) TEM wave sees the same environment it was optimized for is not the prevailing part anymore (Fig 3.5). On the other hand, if the ground separation is reduced, the performance of the single unit cell and therefore the performance of the entire DT will be diminished as can be seen in Fig. 3.4.

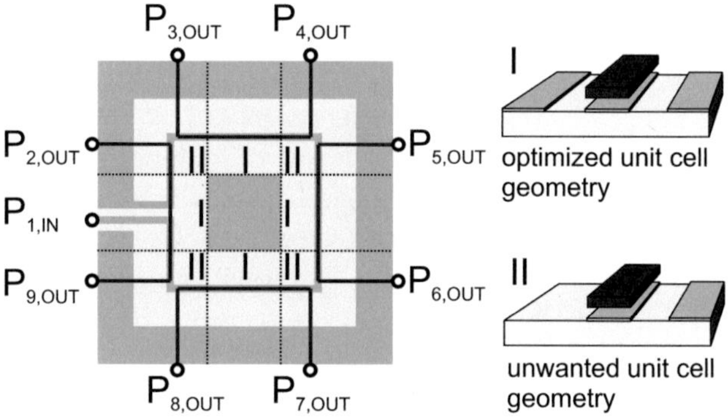

Figure 3.5: The influence of the circular DT geometry on the unit cell environment.

The two main advantages of the circular transformer design are the availability of virtual grounds (always two adjacent ports build a differential port), and the compactness of the overall transformer. However, if it is not for DATs, where the active elements are directly connected to the transformer, the compactness is actually no advantage at all, because the output or input ports of the applied gain cells (more complex circuits than single transistors) have to be connected to the ports of the transformer and usually they do not fit directly next to the DT. If lines are used to connect the gain cells with the transformer, their lengths have to be identical to preserve the phase relation between the various ports on either side of the lines. In the case of a circular geometry, the various ports of the transformer and the gain cells usually do not have the same distance from each other, which means that the lines have to be quite long and interwound. This adds unwanted loss and requires costly wafer space. Therefore, if it is not necessary to have always two out of phase ports close to each other (one differential port), e.g. if efficient ground connections are available, the circular geometry is actually not the best possible geometry for a DT as part of an IMN/OMN. A far more appropriate configuration is a straight connection of all unit cells (Fig. 3.6), as it takes almost no influence on the behavior of the single unit cells and even supports the distribution of signals by providing ports that are further apart than for a circular geometry. Other than that, a straight DT can provide all ports on the same side, which eases the problem of connecting the DT with the gain cells further.

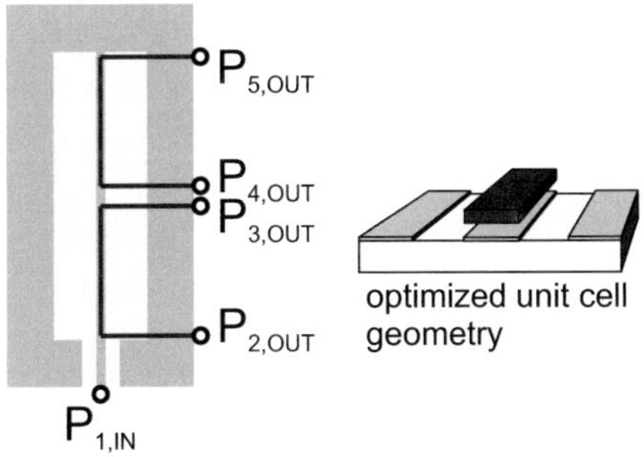

Figure 3.6: The straight DT geometry.

If the separation between some of the single low impedance ports in the straight configuration is not sufficient, low impedance lines have to be used to connect the DT to its destination (IMN) or origin (OMN). Low impedance lines however are always quite lossy [PDK+14], which is why their length should be minimized, if possible. To do this, slight bends can be inserted between unit cells, whose ports are too close to each other to connect two neighboring gain cells directly to both of them (e.g. Fig. 3.6 port three and four). An example is depicted in Fig. 3.7. Depending on the ground separation of the unit cells, already a very small angle of the bend can lead to a notable increase of the separation between the ports, with almost no influence on the single unit cell's environment, guaranteeing the correct response of the overall DT.

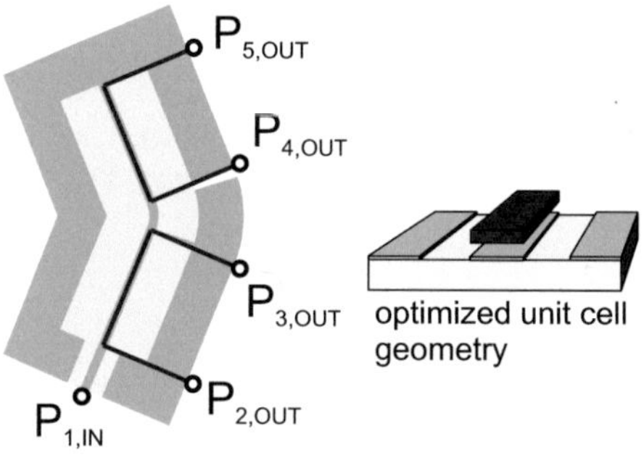

Figure 3.7: The bended DT geometry.

In the next section EM simulations of complete DT based IMNs will be demonstrated. The DTs will be built by the asymmetrical, broadside coupled unit cell proposed in the previous section, which provided the best properties among all investigated unit cell geometries. To minimize the deviation between the behavior of a single unit cell and the complete DT, the unit cells will be connected in the straight DT configuration, proposed in this section.

3.5 Implementation and realization

Now that the geometry of the unit cell as well as the DT has been determined based on EM simulations and the requirements for the IMN/OMN, the performance of the entire DT shall be demonstrated in this section. Fig. 3.8 shows the schematic and the top view of the actual 5 port DT layout, that will be evaluated. Even though, all DTs can be used as IMN as well as OMN, the section ahead will again assume that the designed DT is for an IMN, which defines which side of the DT is the input and output side.

In most IMNs it is convenient to have the input port of the network (P_1) on one and all output ports ($P_2 - P_5$) on the opposite side of the IMN. The input port of the IMN will be usually directly connected to an RF pad and therefore is located at the edge of an MMIC. On the IAF process RF pads end in co-planar waveguide (CPW) lines with either 50 µm or 14 µm gnd-to-gnd separation. At high mmW frequencies, unwanted substrate modes can propagate on cpw lines with 50 µm gnd-to-gnd separation, which is why usually CPW lines with 14 µm are used at these frequencies. A 14 µm CPW line consists only of Met1, which is why the primary, i.e. single-ended side of the DT is implemented in Met1 as well. After a short 14 µm CPW line of 50 Ω, which is merely long enough to support the correct propagation mode, the input signal continues on a 50 Ω slot line as soon as it enters the domain of the DT. The first relevant DT element, the input signal passes, is the capacitor on the primary side (C_p), which is implemented as a Metal Insulator Metal (MIM) capacitor (Fig. 3.2 - e). After the primary capacitor, the input signal runs through the four asymmetric DT unit cells before it is shorted to gnd at the end of the last coupled line section. The input signal on the Met1 line (primary side) induces output signals on the four MetG lines (secondary side). The secondary sides of the couped lines end directly on the DT's secondary side capacitors (C_s), which are implemented as MIM capacitors, too. After these capacitors, usually short low impedance lines follow, which is why the DT ends in four very short air microstrip (airMS) lines, which, among all standard lines available on the Fraunhofer IAF process, provide the lowest characteristic impedance.

The layout parameters, which are determined by the optimization results for the single DT unit cells, are indicated in Fig. 3.9. Based on the evaluation of the various possible unit cell geometries for the asymmetrical coupled line configuration, a suitable geometry in terms of *cwidth* and *gnd_sep* is selected first. By choosing the coupled line geometry, the coupling factor is fixed. Depending

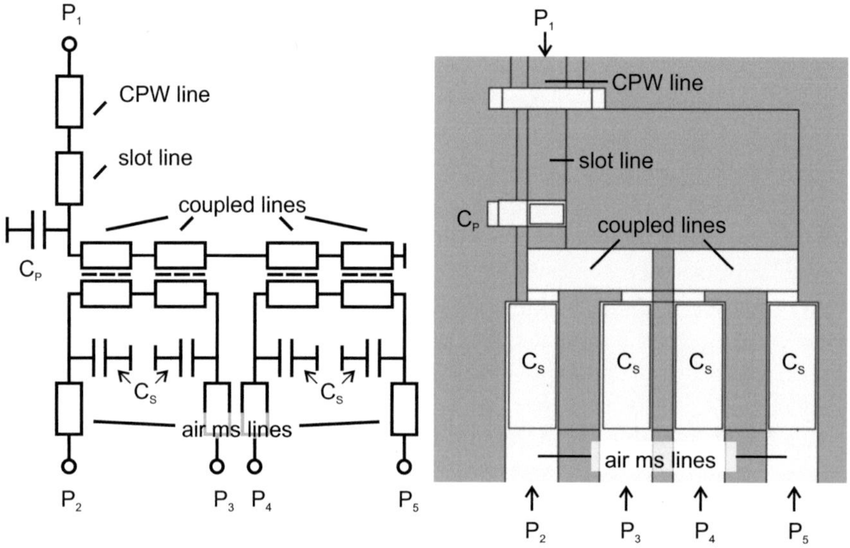

(a) Schematic of a 5 port DT.

(b) Top view of one possible layout for a 5 port DT on IAF's mHEMT MMIC technology.

Figure 3.8: Schematic and layout of one possible 5 port DT configuration.

on the DT geometry (four unit cells in series) and the port impedance at port 1 (Z_{P1}), the load termination (Z_0) for the optimization based on the equations of chapter 2 for the single unit cell is determined ($Z_0 = \frac{Z_{P1}}{4}$). For an optimization for maximum efficiency the quality factor is necessary, which can be determined from the Z-parameters of the coupled line for a specified coupled line length (Eqn. 2.20 - $Q_{p,s} = \frac{\omega L_{p,s}}{R_{p,s}} = \frac{\mathrm{Im}\{Z_{pp,ss}\}}{\mathrm{Re}\{Z_{pp,ss}\}}$). Since the inductance and resistance of the primary and secondary side of the transformer depend on the length of the coupled lines, the optimization for the optimum inductance, i.e. the optimum length, for maximum efficiency has to be done iteratively, until the simulated (shorted unit cell element of the length l) and calculated inductance L match. If the loss and therefore the quality factors are neglected in the optimization of the single unit cell, like for the optimizations for maximum symmetrical bandwidth, the appropriate length (l) of each coupled line section can be either determined by the inverse of Eqn. 3.5 or also iteratively.

After defining all the parameters of the coupled line unit cell, the next step is the design of the parallel capacitors C_s and C_p. On the primary side the necessary capacitance can be realized easily, as its value is always small ($C_p \approx \frac{C_s}{\text{number of unit cells}}$). On the secondary side however the value is not so small, which makes it necessary to use wider MIM capacitors to guarantee that the capacitor is operated well below its self resonance frequency. As visible in Fig. 3.9(a), wider capacitors influence the electromagnetic environment of the unit cell more and due to the limited length of the coupled line sections their maximum width (at least in a DT with a straight unit cell cascadation) is limited. Therefore, a trade-off between length and width is inevitable, which makes it necessary to model the MIM capacitor on the secondary side as a low characteristic impedance line to determine its length from the capacitance value obtained from the optimization equations.

One problem that shall be mentioned, is the relatively small difference between the load termination at the output port ($Z_{out} = \frac{Z_{in}}{\text{number of unit cells}}$) and the characteristic impedance of the low impedance line, resembling C_s, which as the difference gets smaller, leads to a severe deviation between the behavior of a lumped capacitor and its actual implementation as low impedance line. Despite this issue, in general an acceptable trade-off between width and length of the parallel capacitor can be found.

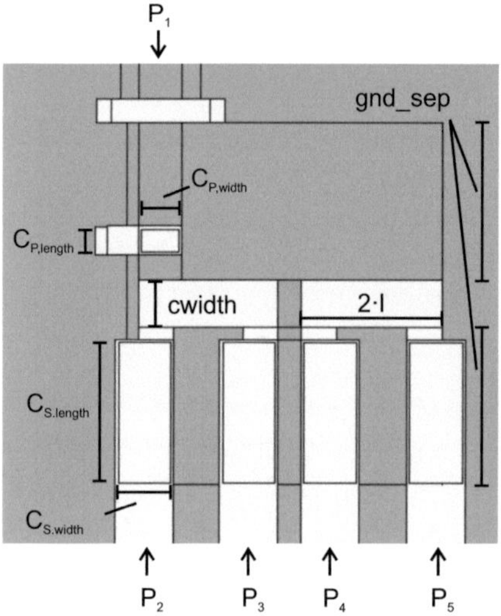

(a) Top view of one possible layout of a 5 port DT with annotated geometry parameters.

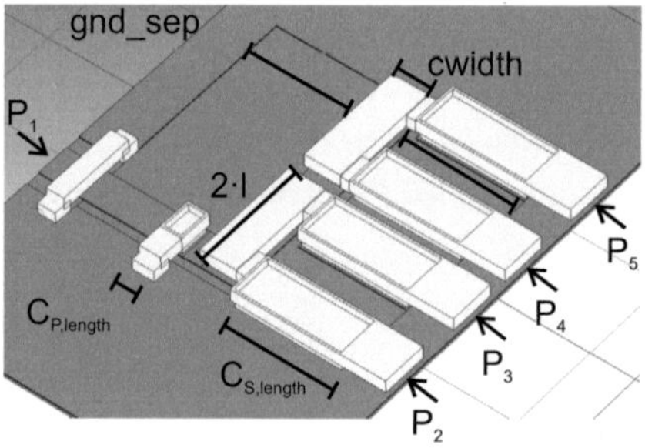

(b) Perspective view of one possible layout of a 5 port DT with annotated geometry parameters.

Figure 3.9: Layout of one possible 5 port DT configuration with annotated geometry parameters.

In order to select the best unit cell geometry for the IAF process, in terms of

- efficiency (η)
- symmetrical bandwidth
- amplitude imbalance (ΔA)
- phase imbalance ($\Delta\phi$)

the unit cell's geometry parameters can be swept in EM simulation to attain S-parameter files for all possible configurations. Based on these S-parameter files and using the new formulas derived in chapter 2, the optimization for maximum efficiency or maximally flat frequency response, i.e. symmetrical bandwidth, at 240 GHz can be performed for all geometries, based on the extracted per unit length quantities ($\mathrm{R}', \mathrm{L}', \mathrm{G}', \mathrm{C}'$). The results of this optimization are the basis for finding an acceptable trade-off between the DT properties.

3.5.1 Maximum efficiency optimization

In Fig. 3.10 the results for an optimization for maximal efficiency, based on the S-parameter files of all geometries are depicted versus the geometry variables *ground_sep* and *cwidth*. C_p and C_s are ideal. All important DT parameters, listed in the previous section are shown. To determine these parameters without doing numerous EM simulations of the entire DT, the behavior of the DT was evaluated by combining the S-parameters of four DT unit cells to get the S-parameters of the DT. That this is a valid approach to determine the best trade-off between different DT attributes will be proven later in this section.

In Fig. 3.10(a) and 3.10(b) the efficiency of the 5 port DT and that of a single DT unit cell are plotted. As visible, the efficiency of the overall network is lower than that of a single unit cell element of the 5 port. For small *cwidth* values the efficiency is almost identical, but for wider lines of the unit cell, the efficiency of the DT drops notably compared to the unit cell. As the coupled lines get wider, the magnetic coupling coefficient (Fig. 3.10(i)) increases as well. This, as seen in the previous chapter, leads to an increase in the optimal transformer inductance. For a given coupled line unit cell geometry, a higher inductance leads to a longer coupled line (Fig. 3.10(g), 3.10(h)). A longer unit cell, however results in a more severe deviation between the phases (Fig. 3.10(f)) and amplitudes (Fig. 3.10(c)) of the various DT ports, which is the main reason for the efficiency drop for wider coupled lines. Besides the width of the coupled lines (*cwidth*) the efficiency of both, the unit cell and the overall DT also varies with the ground separation and the efficiency of the DT peaks for a ground separation of about 20 μm. The reason for this is partly the coupling coefficient again, with its influence on the DT's port phases. Moreover, the quality factors of the two coupled lines play a role. In Fig. 3.10(j) and 3.10(k) the quality factors of the primary (Met1) and the secondary (MetG) side of the DT unit cell are depicted. As expected the secondary side has a higer quality factor, as it is implemented by the galvanic metal layer (MetG), which has a higher conductivity and is also thicker (table 3.2) than Met1. In Fig. 3.10(l) the mutual quality factor ($Q_m = \frac{\omega M}{R_m}$) is depicted. Note that unlike for the quality factors Q_p and Q_s, a lower Q_m leads to a higher efficiency because the mutual resistive coupling increases.

The relative symmetrical bandwidth of the efficiency optimized unit cell (Fig. 3.10(e)) and the 5 port DT (Fig. 3.10(d)) is almost identical. Here, the bandwidth is defined as the frequency range in which the RL is more than 10 dB symmetrical around the optimization frequency of 240 GHz, simulated at the input port (port 1) of the IMN.

To verify the significance of the results of Fig. 3.10 for an entire DT, which is set up as shown in Fig. 3.8, the DT simulations based on S-parameter files of the coupled lines unit cells and ideal capacitors shall be compared with full EM simulations of the entire DT next.

First example of a 5 port DT optimized for maximum efficiency at 240 GHz

According to Fig. 3.10(a), the first configuration listed in table 3.6 should result in one of the highest possible efficiencies for the DT.

Since the simulations for the optimization based on the S-parameters of the coupled line unit cell geometries incorporate solely the losses of the coupled lines and neglect all additional losses due to connecting lines and lossy parallel capacitors, the actual full wave EM simulation of the entire DT is expected to have

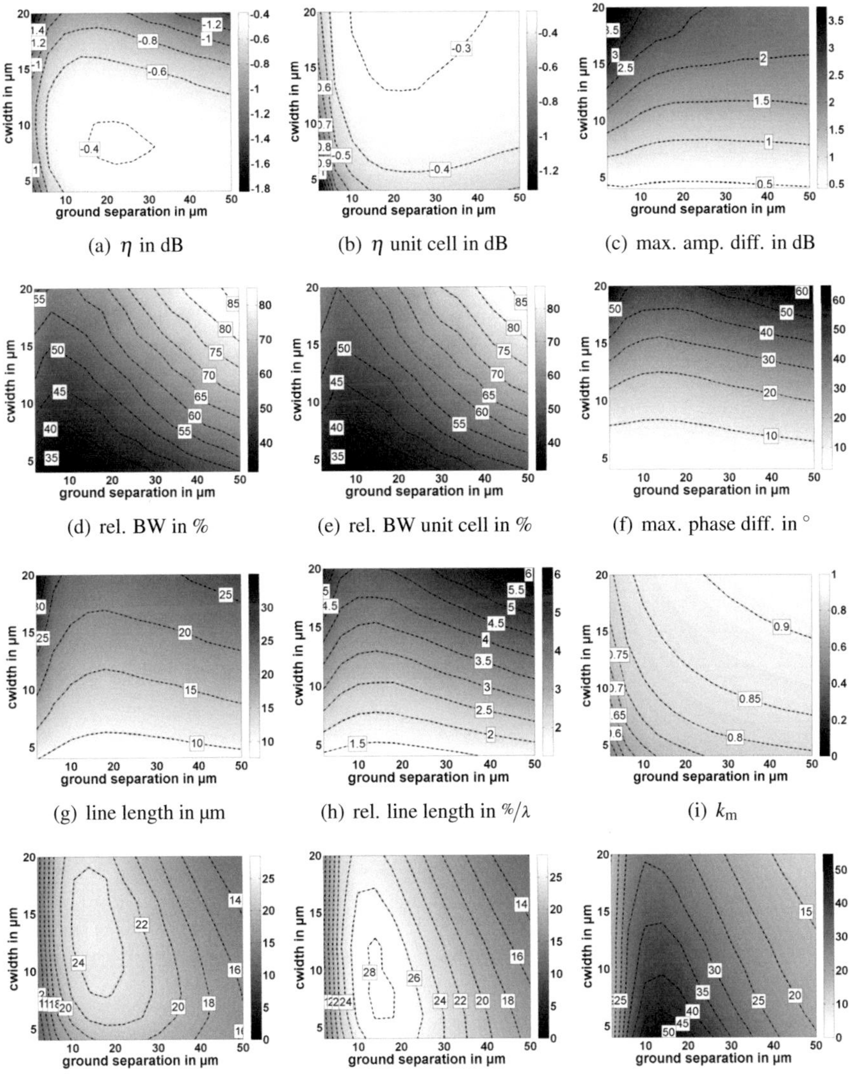

(a) η in dB (b) η unit cell in dB (c) max. amp. diff. in dB

(d) rel. BW in % (e) rel. BW unit cell in % (f) max. phase diff. in °

(g) line length in µm (h) rel. line length in %/λ (i) k_m

(j) Q_p (k) Q_s (l) Q_m

Figure 3.10: Optimization results for optimal efficiency of a 5 port DT using the IAF 50 nm and 35 nm processes vs. *cwidth* and *gnd_sep* at 240 GHz.

more loss and therefore to be less efficient than what is anticipated by Fig. 3.10. However, the efficiency of the DT should still peak at the design frequency of 240 GHz.

cwidth	8 µm
gnd_sep	26 µm
l	12 µm
cp_width	7.4 µm
cp_length	3.6 µm
cs_width	10 µm
cs_length	23.4 µm

Table 3.6: Geometry of the first investigated 5 port DT optimized for maximum efficiency.

In Fig. 3.11 the reflection and transmission coefficients of the complete 5 port DT determined from the cascadation of optimized unit cells and a full wave EM simulation are compared for port 1 and 6.
Here, port 6 is no physical port, as the DT has only 5 ports in total, but stands for the sum port of an ideal power combiner, that combines all output ports (P_2 - P_5) to just one port (P_6) (Fig. 3.12).
The S-parameter matrix of the ideal coupler is given in Eqn. 3.8 and its rows and columns are chosen to fit the port numbers shown in Fig. 3.12.

$$\left[S_{ideal4:1}\right] = \begin{bmatrix} 0 & 0 & 0 & 0 & 0 & 0 \\ 0 & 0 & 0 & 0 & 0 & \frac{1}{2} \\ 0 & 0 & 0 & 0 & 0 & -\frac{1}{2} \\ 0 & 0 & 0 & 0 & 0 & \frac{1}{2} \\ 0 & 0 & 0 & 0 & 0 & -\frac{1}{2} \\ 0 & \frac{1}{2} & -\frac{1}{2} & \frac{1}{2} & -\frac{1}{2} & 0 \end{bmatrix} \tag{3.8}$$

The comparison of the two simulation results demonstrates that both reflection coefficients show a good agreement at the design frequency. Below 240 GHz the additional lines in the EM simulation result in a visible line transformation

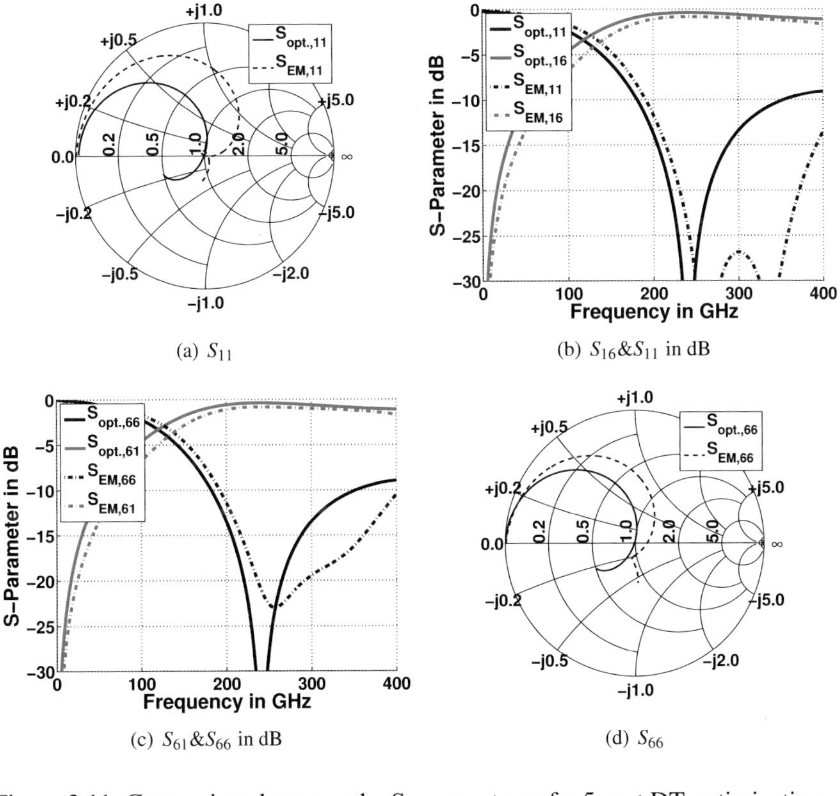

(a) S_{11} (b) S_{16}&S_{11} in dB

(c) S_{61}&S_{66} in dB (d) S_{66}

Figure 3.11: Comparison between the S-parameters of a 5 port DT optimization result for optimum efficiency at 240 GHz and a complete EM simulation. While port 1 represents the real port 1 of the DT, port 6 represents the ideally combined ports 2 to 5.

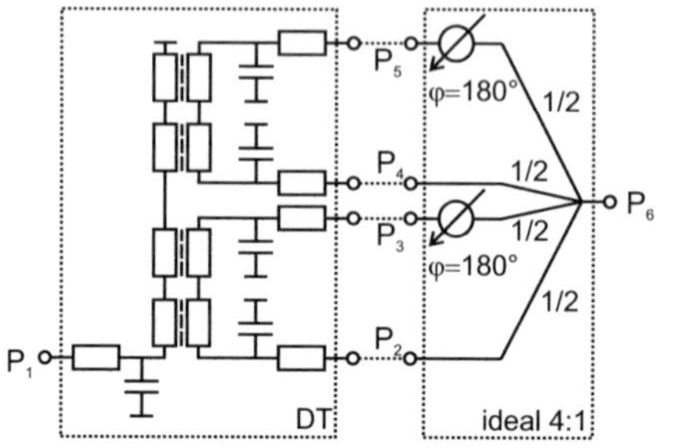

Figure 3.12: Schematic of the DT in combination with the ideal combiner. All ports of the DT and combiner are indicated.

along the m-circle. Above the design frequency, the EM simulation shows a wider bandwidth than the optimization based on the cascaded unit cells plus ideal capacitors. The reason for this is most likely the distributed behavior of the parallel capacitors in the EM simulation and the non-ideal electrical short at the end of the last coupled line element as well as the inductive transition from the strip line to the coupled line section. In the optimization the electrical short at the end of the last coupled line section is assumed to be ideal. In the real implementation the electrical short at the end of the primary side is slightly inductive, due to the non-negligible ground_sep. The same is true for the transition from the strip line to the coupled line section, which is neglected in the optimization, too. However, in general the reflection and transmission coefficients of both the EM simulation and the optimization match well.

In Fig. 3.13 the maximum amplitude and phase variation between all four output ports (P_2-P_5) is plotted. Here, the 180° phase shift between the various ports has already been deducted. Around the optimization frequency a good agreement between the EM simulation and the optimization result is visible for the amplitude as well as the phase variation.

To verify that the combination of coupled lines and parallel capacitors results in the highest efficiency for the given losses of the single elements, Fig. 3.14 plots the transmission coefficient (S_{61}), efficiency (η) and the maximum available

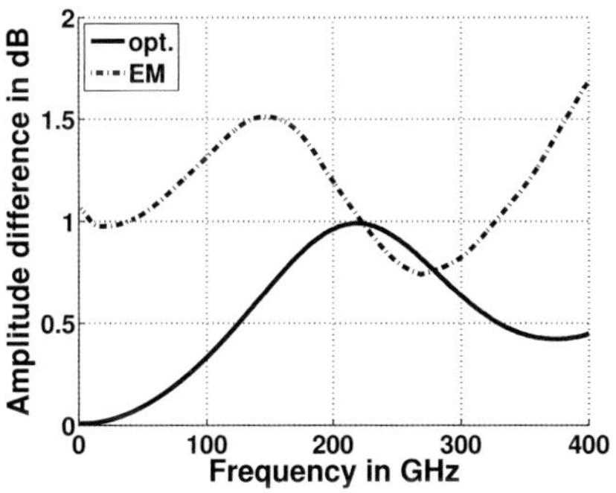

(a) Maximum amplitude difference in dB between the ports 2 to 5 for a signal injected at port 1.

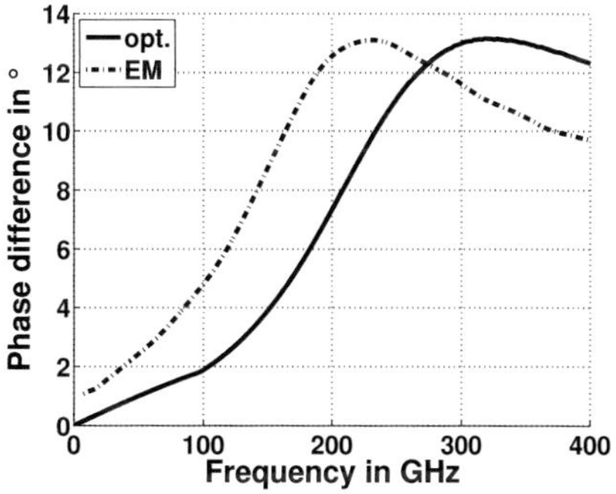

(b) Maximum phase difference in degree between the ports 2 to 5 for a signal injected at port 1.

Figure 3.13: Comparison between the maximum amplitude and phase difference of a 5 port DT optimization result for optimum efficiency at 240 GHz and complete EM simulation of the structure.

gain (MAG) for the optimization simulation as well as the EM simulation of the entire DT.

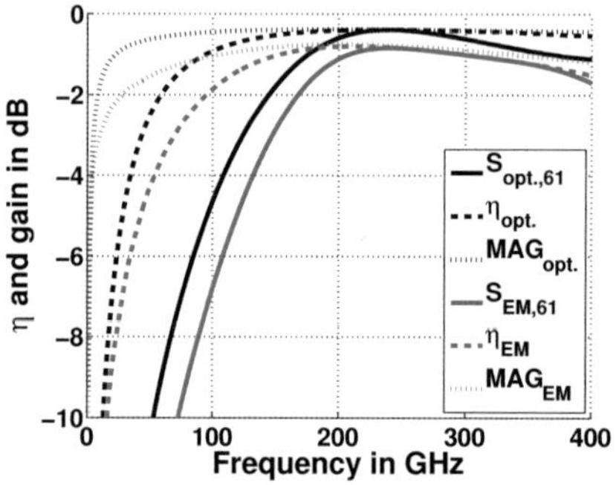

Figure 3.14: Transmission coefficient, efficiency and MAG of a 5 port DT optimization result for optimum efficiency at 240 GHz and an EM simulation of the complete DT.

Note that for each of the simulations the transmission coefficient, the efficiency and the MAG meet at 240 GHz, which means that the optimization for maximum efficiency, based on the equations of the previous chapter, is successful. The variation between the two simulations shown in the figure is due to the losses of the parallel capacitors and additional line elements, which are neglected in the unit cell optimization simulation. While the simulation based on the coupled line sections and ideal capacitors predicts an IL of 0.38 dB at 240 GHz the EM simulation of the entire DT (Fig. 3.8) has an IL of 0.77 dB, which is about twice the loss.

As depicted in Fig. 3.13(a) and 3.13(b), the maximum amplitude and phase imbalance between the four output ports are about 1 dB and 12° respectively at 240 GHz. If these values are too high for the intended application, a better amplitude and phase balance can be traded for a slightly worse efficiency by picking a different coupled line unit cell.

Second example of a 5 port DT optimized for maximum efficiency at 240 GHz

In table 3.7 another set of DT geometry parameters is listed. This geometry is also the result of an optimization for maximum efficiency as the one before. This time, however, a coupled line geometry is chosen that should result in a better amplitude and phase balance. By selecting a geometry that has thinner lines and a wider ground separation, the mean value of the weighted characteristic impedance of the asymmetrical coupled line is increased, which compared to the first example (table 3.6) results in a shorter line (Eqn. 3.5) and therefore, as discussed earlier, in a better amplitude and phase balance.

cwidth	4 µm
gnd_sep	50 µm
l	9 µm
cp_width	7.4 µm
cp_length	5 µm
cs_width	6 µm
cs_length	47.6 µm

Table 3.7: Geometry of the investigated second DT 5 port optimized for maximum efficiency.

Fig. 3.15 compares again the reflection and transmission coefficient of the optimization simulation based on solely the coupled lines sections and ideal capacitors with that of the completely EM simulated DT. The discussion of the results can be summarized more or less identical to that of the previous geometry. Again, a line transformation is visible in the reflection coefficient of port 1 and port 6 and the transmission coefficient is very similar for both simulations. Compared to the responses of the previous configuration, the simulation of the complete DT tends to deviate slightly more from that of the optimization simulation over frequency. This is due to the higher, frequency depended loss in the parallel capacitors on the secondary side for this configuration. As can be seen in table 3.7, the parallel capacitors are narrower and longer than that of the previous DT geometry. Hence, the capacitors are more lossy.

This behavior is also visible in Fig. 3.16, where again the transmission coefficient, efficiency (η) and MAG are plotted for the optimization (opt.) and the full wave EM simulation (EM). Note again, that for both simulations the three curves (transmission coefficient, η and MAG) are identical at 240 GHz, the design frequency of the DT. In this configuration however, the difference between both simulations is more severe, due to the much higher losses in the parallel capacitors and longer connection lines. The long parallel capacitors and lines are unavoidable in this layout configuration, due to the higher ground separation of the DT unit cell (ground_sep) (table 3.7), which has to be bridged to connect the DT to its destination. The IL of the optimum simulation is 0.52 dB, while that of the complete DT EM simulation is 1.22 dB at 240 GHz.
In Fig. 3.17 the maximum amplitude and phase imbalance between all four output ports of the DT is shown. By comparing the curves with those of Fig. 3.13 the expected improvement due to the deliberate change of the unit cell geometry is clearly visible. Both, the amplitude and the phase imbalance dropped from about 1 dB and 12° down to approximately 0.5 dB and 8°, respectively. On the other hand, compared to the previous example, the IL went from 0.77 dB to 1.22 dB in the EM simulation of the DT.
Based on the EM simulation of various geometries of small coupled line DT unit cells, the optimal geometry for maximum efficiency can be chosen. Depending on the acceptable amplitude and phase imbalance, the best geometry can be selected and implemented in a complete DT layout as the one shown in Fig. 3.8. The agreement between the simple and fast simulation and optimization of the entire DT based on the per unit length parameters of a coupled lines section, and the EM simulation of the whole DT is very good. To verify that this is also true for the maximum symmetrical BW optimization presented in chapter 2, the next section will repeat the investigation for a unit cell geometry optimized for maximally flat IL BW around 240 GHz.

3.5.2 Maximum bandwidth and flat IL optimization

As in the previous paragraph for an optimization for maximum efficiency at 240 GHz, Fig. 3.18 shows the important parameters of a 5 port DT built out of

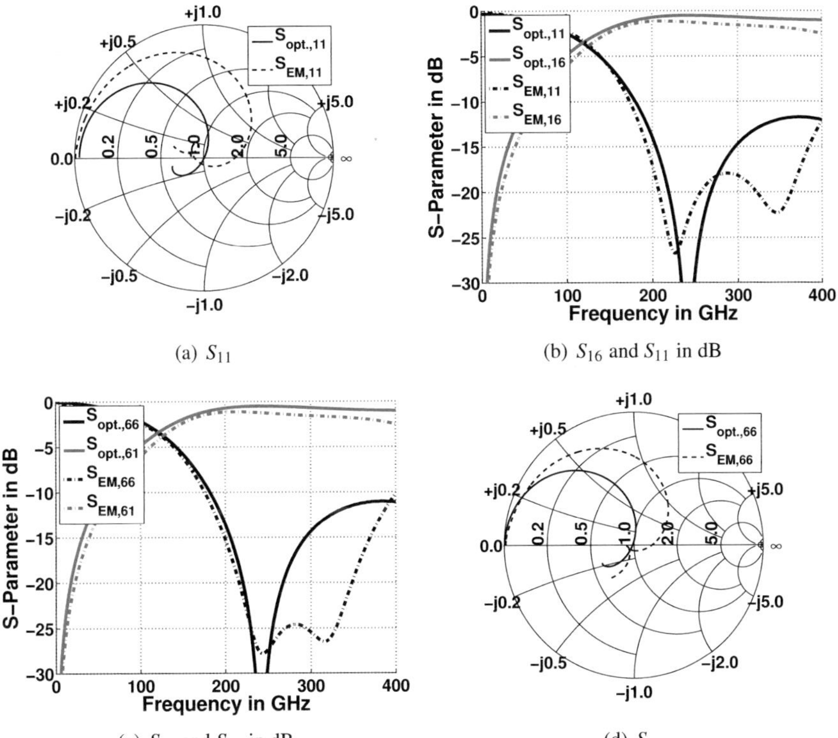

(a) S_{11}

(b) S_{16} and S_{11} in dB

(c) S_{61} and S_{66} in dB

(d) S_{66}

Figure 3.15: Comparison between the S-parameters of a 5 port DT optimization result for optimum efficiency at 240 GHz and an EM simulation of the complete DT. While port 1 represents the real port 1 of the DT, port 6 represents the ideally combined ports 2 to 5.

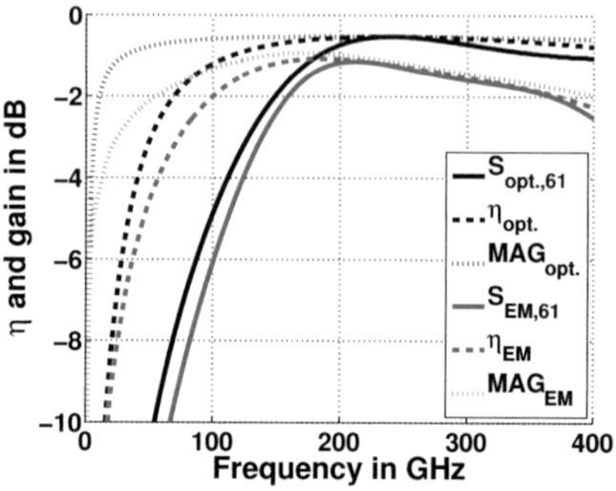

Figure 3.16: Transmission coefficient, efficiency and MAG of a 5 port DT optimization result for optimum efficiency at 240 GHz and an EM simulation of the complete DT.

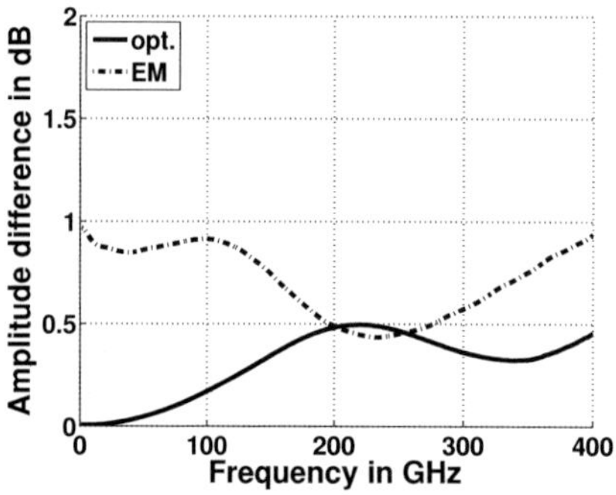

(a) Maximum amplitude difference in dB between the ports 2 to 5 for signal injected at port 1.

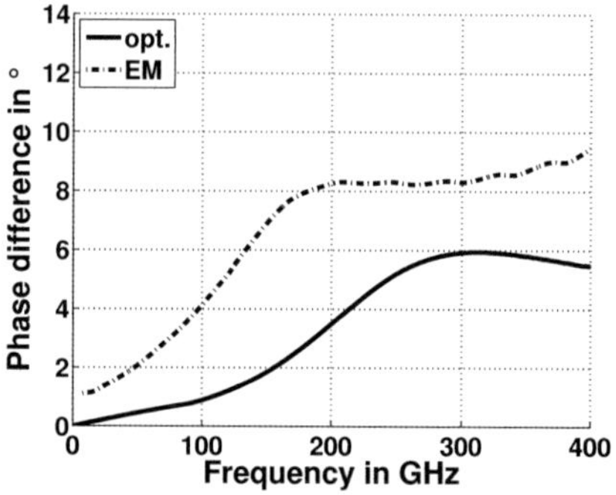

(b) Maximum phase difference in degree between the ports 2 to 5 for a signal injected at port 1.

Figure 3.17: Comparison between the maximum amplitude and phase difference of a 5 port DT optimization result for optimum efficiency at 240 GHz and an EM simulation of the complete DT.

coupled line unit cells and ideal parallel capacitors optimized for a maximally flat frequency response at 240 GHz.

For the same coupled line geometries, the optimization for maximally flat response results in longer coupled line DT unit cells (Fig. 3.18(h)) than that optimized for maximum efficiency (Fig. 3.10(h)). This behavior was already discussed in chapter 2. Due to the longer lines the issue of amplitude (Fig. 3.18(l)) and phase (Fig. 3.18(f)) imbalance is more severe for a DT optimized for a broad symmetrical bandwidth. Amplitude and phase imbalances well below 1 dB and 10° are not possible for this optimization strategy as they would require narrower coupled lines than the lower bound of the process design rules (table 3.1 - MetG width $\geq$ 4 µm) allows.
The maximum efficiency value for this unit cell design strategy is about -0.7 dB and it also occurs for narrow coupled lines (Fig. 3.18(a)). The same is true for the relative symmetrical bandwidth (Fig. 3.18(d)), which peaks at $cwidth = 7$ µm and $gnd\_sep = 50$ µm. As expected for an optimization for maximum symmetrical bandwidth, the relative bandwidth of this optimization strategy exceeds that for maximum efficiency by far. While DTs optimized for maximum efficiency achieve relative bandwidths of no more than 60 %, the optimization for maximally flat symmetrical bandwidth achieves about 100 %. Recall that the bandwidth is defined in terms of the symmetric -10 dB bandwidth simulated at the input port (P_1). Note that unlike for the optimization for maximum efficiency, this time the bandwidth of the single DT unit cell (Fig. 3.18(e)) and that of the complete DT (Fig. 3.18(d)) deviate. Like for the efficiency of the single DT unit cell (Fig. 3.18(b)) and the 5 port DT (Fig. 3.18(a)), the phase and amplitude imbalance are the reason for this.

To verify that the results of the maximally flat response optimization based on S-parameter files of the various coupled line unit cell geometries correspond to the actual results of the complete EM simulated DT, one of the optimization geometries for a broadband coupled line DT will be compared to an EM simulation of a complete DT next. The principle layout of the DT used for the EM simulation is again like the one shown in Fig. 3.8.

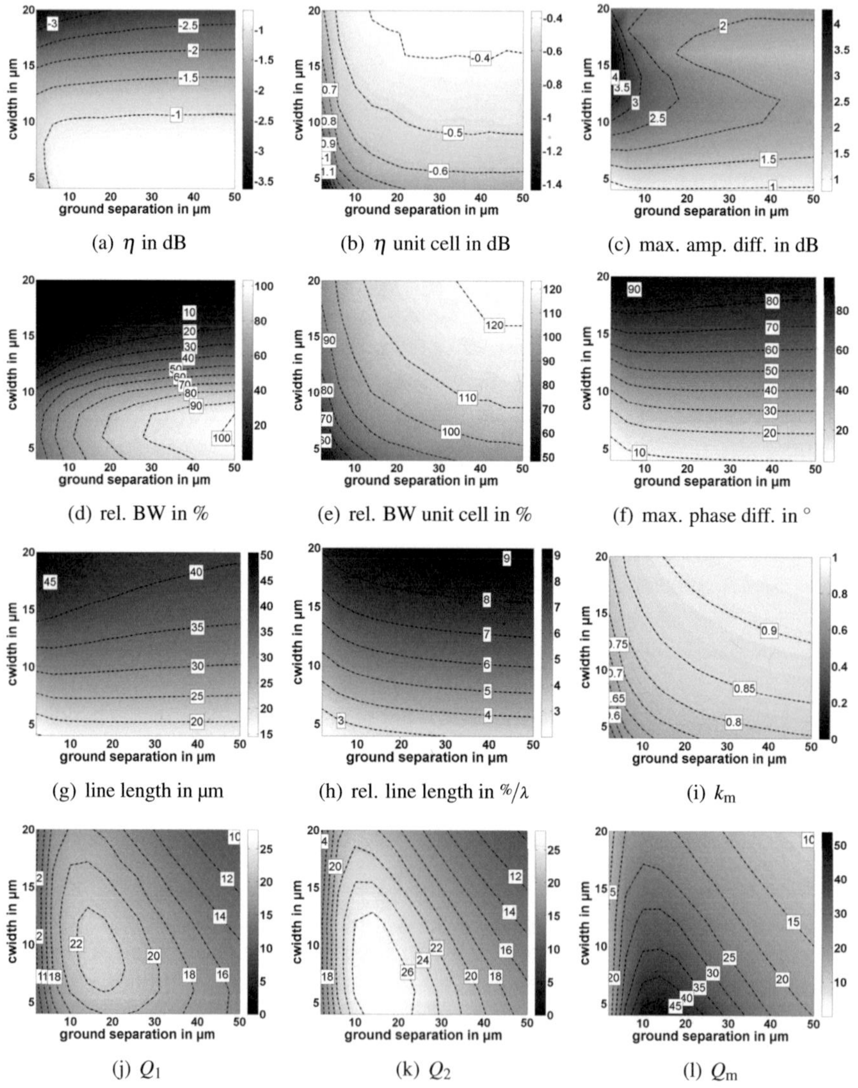

(a) η in dB (b) η unit cell in dB (c) max. amp. diff. in dB

(d) rel. BW in % (e) rel. BW unit cell in % (f) max. phase diff. in °

(g) line length in µm (h) rel. line length in %/λ (i) k_{m}

(j) Q_1 (k) Q_2 (l) Q_{m}

Figure 3.18: Optimization results for a maximally flat frequency response of a 5 port DT using the IAF 50 nm and 35 nm processes vs. *cwidth* and ground separation (*gnd_sep*) at 240 GHz.

Example of a 5 port DT optimized for maximally flat insertion loss, i.e. maximum symmetrical bandwidth, at 240 GHz

The selected unit cell geometry is listed in table 3.8. Besides its length, the coupled line unit cell geometry is identical to that of the second unit cell geometry optimized for maximum efficiency. Due to the fact that the DT is almost twice as long as the previous DT, the width of the capacitors on the secondary side could also be doubled without interfering with the unit cell geometry too much. This makes sure that the loss of the parallel capacitors do not dominate the overall loss of the DT, as it was the case in the second of the two DTs optimized for maximum efficiency.

cwidth	4 µm
gnd_sep	50 µm
l	17.29 µm
cp_width	7.4 µm
cp_length	5.6 µm
cs_width	12 µm
cs_length	19.55 µm

Table 3.8: Geometry of the investigated 5 port DT optimized for maximally flat IL and symmetrical BW.

Fig. 3.19 compares again the reflection and transmission coefficient of the optimization simulation and the EM simulation of the entire DT. A good agreement is visible. The only difference that can be noted is the slightly lower overall bandwidth of the EM simulation of the entire DT. At the upper edge, the passband ends a little bit earlier than what is predicted by the optimization simulation based only on the optimized DT unit cells and ideal capacitors.
The good agreement between both simulations proves that the simulation of DTs based on optimized coupled lines in combination with ideal capacitors (i.e. the DT unit cell) already encompasses the whole relevant frequency behavior and therefore models the DT correctly.

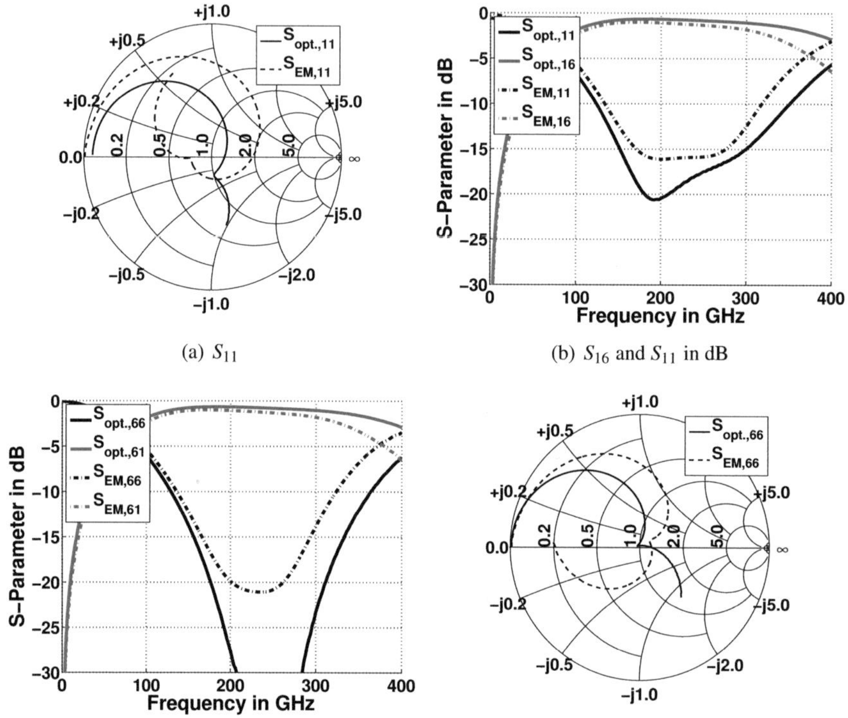

(a) S_{11}

(b) S_{16} and S_{11} in dB

(c) S_{61} and S_{66} in dB

(d) S_{66}

Figure 3.19: Comparison between the S-parameters of a 5 port DT optimization result for maximally flat BW response at 240 GHz and an EM simulation of the complete DT. While port 1 represents the real port 1 of the DT, port 6 represents the ideally combined ports 2 to 5.

In Fig. 3.20, the amplitude and phase imbalance versus frequency is depicted. As for the optimization for maximum efficiency, a good agreement between the optimization simulation and the full EM simulation is visible.

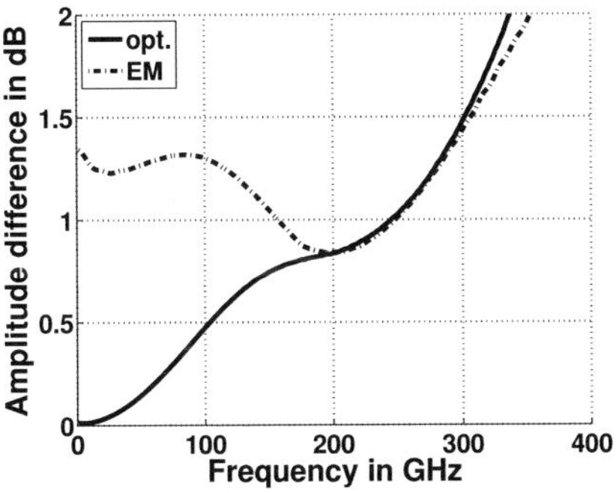

(a) Maximum amplitude difference in dB between the ports 2 to 5 for an input signal at port 1.

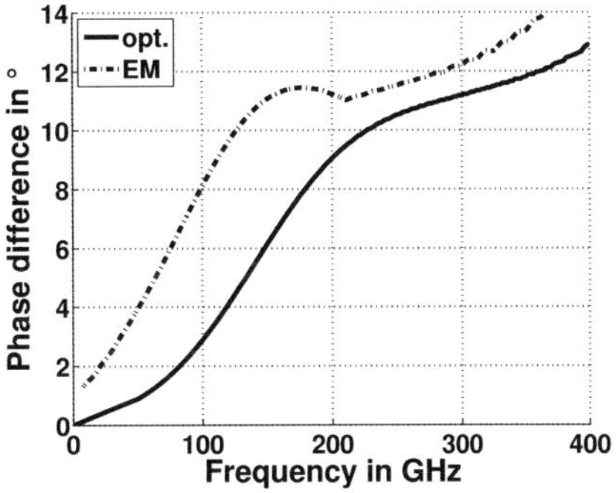

(b) Maximum phase difference in degree between the ports 2 to 5 for an input signal at port 1.

Figure 3.20: Comparison between the maximum amplitude and phase difference of a 5 port DT optimization result for maximally flat bandwidth response at 240 GHz and an EM simulation of the complete DT.

The last curves that shall be compared here, are again the transmission coefficient (S_{61}), efficiency (η) and the MAG. All three curves are plotted for both simulations in Fig. 3.21. This time, the curves of S_{61}, η and MAG do not meet at 240 GHz, since the optimization goal was not the highest possible efficiency but the widest symmetrical bandwidth with a flat response. As expected the response is not really flat, which is because of the negligence of the loss during the optimization. The loss results in a slope towards the upper edge of the passband of the DT. At 240 GHz the optimization simulation predicts an insertion loss of 0.78 dB and the EM simulation assumes about 1.27 dB for the entire DT.

Up to now all investigations regarding the DT have been based either on lumped ECMs or EM simulations. So far, this chapter has proven that the lumped element approach proposed in chapter 2 is valid, since the optimization algorithms for

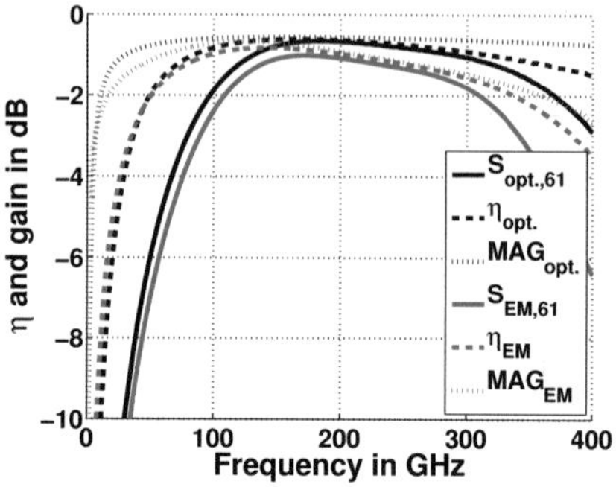

Figure 3.21: Transmission coefficient, efficiency and MAG of a 5 port DT optimization result for maximum flat bandwidth response at 240 GHz and an EM simulation of the complete DT.

maximum efficiency and maximally flat bandwidth proposed in chapter 2 turned out to achieve their objectives in EM simulations of complete DT layouts. The geometry of these DTs abides all the design rules of the applied MMIC technology which means that they can actually be manufactured in the same way they have been simulated in the EM solver. If the EM simulations incorporate all process properties correctly, the measured and simulated DT behave identical. To verify this, the next section will compare EM simulation and measurement results for one possible DT configuration.

3.6 Measurement of DT structures

The measurement of a DT is not straight forward, as the impedance on either the input or output side of the transformer in general is not equal to 50 Ω. To measure the phase and amplitude balance of a multiport DT, usually multiple breakouts of the same DT are necessary. Depending on the number of output (IMN) or input ports (OMN), the same amount of breakout circuits has to be fabricated. In those breakouts always one of the ports is connected to an RF-pad at a time and all others are terminated with the correct port termination. In Fig. 3.22 one possible breakout circuit of a DT is depicted.

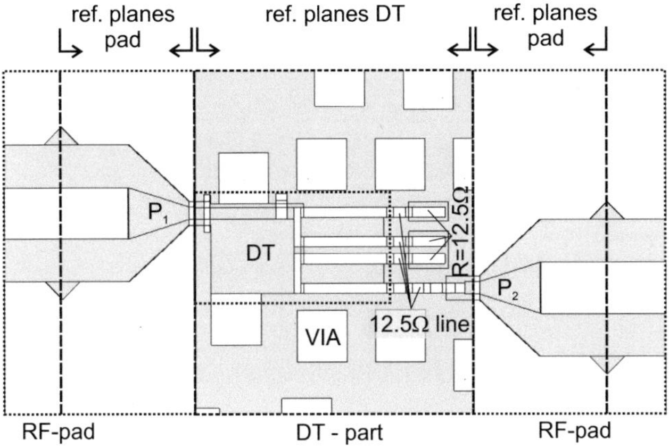

Figure 3.22: Possible breakout circuit for the evaluation of a DT.

On the left and right side of the breakout circuit are RF-pads, which mechanically facilitate the measurement of the circuit. These RF-pads are intentionally designed to behave like 50 Ω lines, because they are usually used as interfaces between RF circuits with 50 Ω reference impedance and 50 Ω RF probes or other 50 Ω circuitry connected by bond wires. At high millimeter-wave frequencies, these 50 Ω RF-pads handicap the measurement of the DT breakout circuits. On one side of the DT the port impedance is usually 50 Ω, which means that the pad does not influence the port matching severely. On the other side of the DT, however, the port impedance is usually well below 50 Ω, which means that the RF-pad results in a line transformation. This line transformation can have a significant influence on the port termination of the DT and lead to an unequal power division between the various output ports of a DT as part of an IMN for instance. In Fig. 3.23 the simulated normalized S-parameters with and without RF-pads for a breakout circuit similar to that depicted in Fig. 3.22 are shown. In these plots the port termination at port 1 (P_1) and 2 (P_2) are 50 Ω and 12.5 Ω, respectively. The

DT used as example in this section is optimized for a symmetrical BW around 240 GHz.

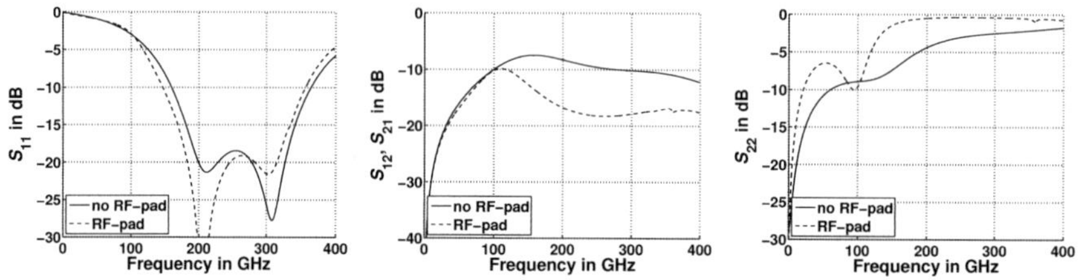

(a) S_{11} in dB with and without RF-pads.

(b) S_{21} in dB with and without RF-pad.

(c) S_{22} in dB with and without RF-pad.

Figure 3.23: Influence of the RF-pads on the analysis of individual DT ports, based on simulations.

The influence of the line termination due to the 50 Ω pad at port 2 of the DT is almost negligible with regard to the matching of port 1. Due to the correct port termination of three of the four output ports (Fig. 3.22), the input port (P_1) is still matched well. The insertion loss between port 1 and port 2, however, suffers severely due to the unwanted line transformation caused by the RF-pad at port 2. In Fig. 3.23(b) the simulated curves of only the breakout circuit and that of the breakout circuit with attached RF-pads differ significantly. The unwanted line transformation increases the IL between the ports because of the mismatch it causes. If there were no losses at all, one would expect an IL of about 6 dB for a 1:4 DT. With the RF-pad attached, the IL between port 1 and port 2 never gets close to that value. On the other hand, the simulation of exactly the same breakout circuit but without any RF-pads, shows a response that is close to what one would expect. Even though the shape is not ideal in terms of DT loss, the response resembles that of a DT, as it was introduced in the previous chapters. The relatively high loss can be traced back to the additional elements in the breakout circuit, which alter the performance of the DT. The last plot of Fig. 3.23, subfigure 3.23(c), compares the reflection coefficient at port 2 of the breakout circuit with and without the RF-pads. Note that both curves show a good port

matching at low frequencies, which is not surprising, as the 12.5 Ω port is *directly* connected to a 12.5 Ω load resistor at those low frequencies. Neither the RF-pad nor the elements of the DT have any influence. All the 12.5 Ω port sees is the 12.5 Ω load resistor. As the frequency goes up, both curves start to get worse. Port 2 does not show any significant matching around the operation frequency of the DT. The reason for this is that port two is part of a differential port. All ports on the differential, i.e. multiport side of a DT need to be excited with their correct phases simultaneously, to show the proper matching behavior.
It can be concluded that, since the RF-pads are indispensable for the measurement of DT breakout circuits, their effects on the measurement have to be compensated afterwards, to get a reliable testimony about the frequency response of the DT. Hence, the compensation of the unwanted line transformations on either side of the DT by de-embedding the RF-pads on both sides of the breakouts will be discussed next.

3.6.1 De-embedding techniques suitable for the mmW regime

Two different approaches have been investigated for the de-embedding of the RF-pads. One is the Open-Short (OS) method [KGV91, TH03] and the other one is the rigorous de-embedding using ABCD parameters of EM simulations of the ports.
The Open-Short method enables the de-embedding of the ports based on measured open-ended and shorted RF-pads. It is based on a simple lumped ECM consisting of a parallel admittance and a series impedance for each port. The Y-parameters [Poz04] of the Device Under Test (DUT) (Y_{DUT}) can be extracted from the Y-parameters of the measured DUT (Y_{M} - including the RF-pads) and the Y-parameters of the open-ended (Y_{O}) and shorted (Y_{S}) RF-pads by evaluating the following equation [LPGG09, Zwi05]:

$$Y_{\mathrm{DUT}} = \left[(Y_{\mathrm{M}} - Y_{\mathrm{O}})^{-1} - (Y_{\mathrm{O}} - Y_{\mathrm{S}})^{-1} \right]^{-1} \tag{3.9}$$

To verify the applicability of the Open-Short method for the de-embedding of the DT's RF-pads, the EM simulated S-parameters of the DT breakout (Fig. 3.22) without the RF-pads are compared to the S-parameters of the entire circuit including the pads after de-embedding of the pads with the Open-Short method. The result is depicted in Fig. 3.24. There, it can be seen that the S-parameters match quite well up to about 110 GHz before the de-embedded S-parameters start to deviate significantly from the true S-parameters of the DUT.

Thus, even though the Open-Short method provides a convenient way to compensate the RF-pads and their influence, based on simple measurements or simulations of an open-ended and shorted RF-pad, it cannot be used for the extraction of the DT S-parameters beyond 110 GHz.

If reliable EM simulations of the RF pad are available, these simulations can be used to extract the S-parameters of the DUT by de-embedding based on ABCD parameters. By converting the S-parameters of the measured or simulated DT breakout circuit into ABCD-parameters, the effects of the RF-pads can be removed by using the following equation:

$$ABCD_{\mathrm{DUT}} = ABCD_{\mathrm{RF\text{-}pad}}^{-1} \cdot ABCD_{\mathrm{M}} \cdot ABCD_{\mathrm{RF\text{-}pad,inv}}^{-1} \quad (3.10)$$

where $\mathrm{ABCD_{DUT}}$, $\mathrm{ABCD_{RF\text{-}pad}}$ and $\mathrm{ABCD_{RF\text{-}pad,inv}}$ represent the ABCD-Parameters of the DUT, the RF-pad and the port inverted RF-pad, respectively.

The S-parameters of the de-embedded DT breakout circuit, based on simulated ABCD-parameters, are also depicted in Fig. 3.24. As expected by this rigorous method, the simulation of the DT with RF-pads after ABCD-parameter de-embedding equals that of the simulated DT without RF-pads, over the entire simulated frequency range. The small visible deviations are due to inaccuracies of the EM simulation.

The ABCD-parameter method does not depend on a lumped ECM to describe the behavior of the RF-pad, like the OS method does, but compensates the true electromagnetic behavior of the RF-pad. It therefore could be expected that the ABCD-parameter based de-embedding method is superior to the OS method. However, at least up to 110 GHz the OS method enables an acceptable de-embedding of the DT based only on simple measurements. No EM simulations are necessary, which can be advantageous.

For the comparison between the OS-method and the ABCD-method, EM simulations of a DT breakout circuit have been used. To excite the structure, waveguide

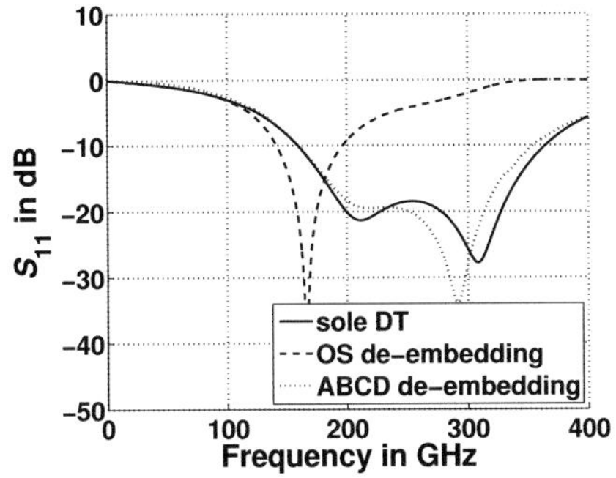

(a) S_{11} in dB

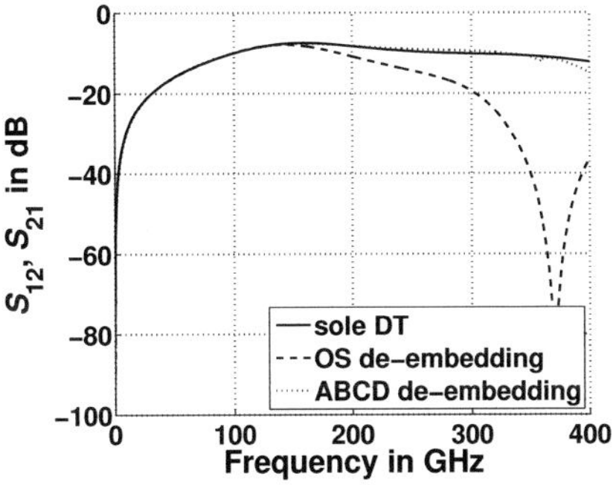

(b) S_{12}, S_{21} in dB

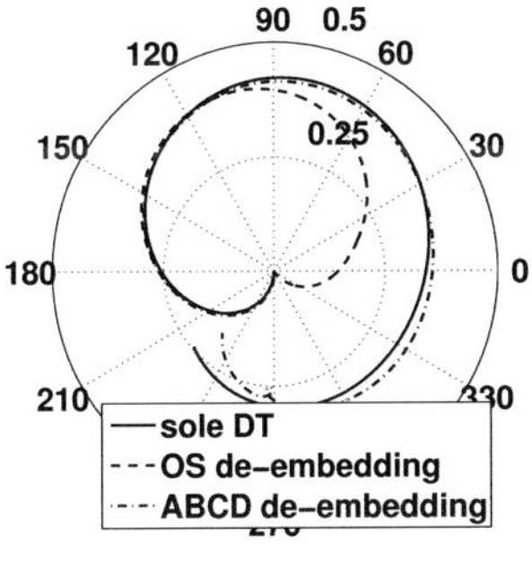

(c) S_{12}, S_{21}

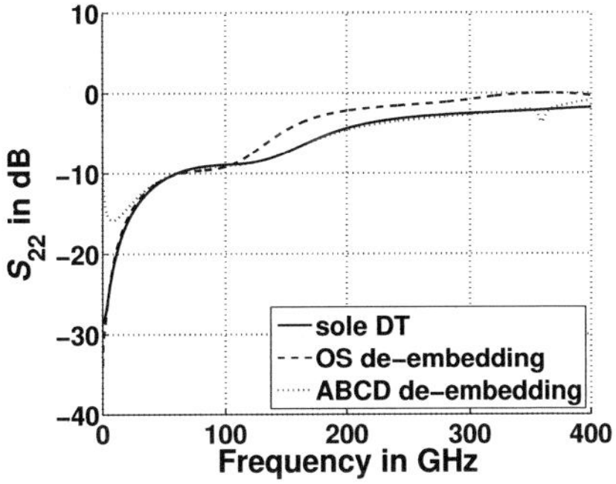

(d) S_{22} in dB

Figure 3.24: Verification of the de-embedding techniques to remove the the RF-pads. The verification is based on EM simulations of a DT with and without an RF-pad and the RF-pad itself.

ports are put at the reference planes (Fig. 3.22) of the RF-pad and the DT part respectively. While the waveguide port at the RF-pad - DT interface models the interface correctly, the waveguide port at the RF-probe - RF-pad interface introduces small errors, since an RF-pad exited by a waveguide port behaves slightly different from that exited by an RF-probe. The behavior of the RF-pad is therefore expected to differ slightly from that of a measured RF-pad. However, the error can be neglected compared to the general problem of accurate probe-tip calibration at high mmW frequencies [WM92] and the imprecision in the positioning of the probe-tip during measurements [DWSE$^+$11, Die14]. In the upcoming investigation a DT operating around 240 GHz will be demonstrated. Hence, the ABCD-parameter based de-embedding method will be used in the subsection ahead. The small errors caused by de-embedding the measurements with EM simulations of the RF-pads that are excited by waveguide ports will be neglected.

3.6.2 Comparison between simulated and measured behaviour of DT structures

Up to now only simulations of various types of DTs have been compared. In this subsection measurement results for a DT operating in the entire G- (140 GHz - 220 GHz) and H-band (220 GHz - 325 GHz) will be presented and compared to measurements.

The principle geometry of the DT is identical to that of the DT simulated throughout the entire chapter and visible for instance in Fig. 3.8.

In table 3.9 the parameters of the manufactured geometry are shown.

Note that even though the parameters of the coupled line (*cwidth*, *gnd_set*) of the unit cell are identical to those optimized for maximum efficiency and maximally flat response in this chapter, the length of the coupled line unit cell does not equal either of these two cases, but lies somewhere between them. The length was chosen as a compromise between a good efficiency and a wide symmetrical bandwidth as well as an acceptable phase and amplitude balance. Therefore, the response of the presented DT is not necessarily ideal with respect to loss or bandwidth but demonstrates the performance of the DT in general.

In Fig. 3.25 a photograph of the manufactured 5 port DT breakout test field, used for the measurements in this section, is shown. Besides a back-to-back config-

cwidth	4 µm
gnd_sep	50 µm
l	11.6 µm
cp_width	6.8 µm
cp_length	5.5 µm
cs_width	6 µm
cs_length	49 µm

Table 3.9: Geometry of the measured 5 port DT.

uration (Fig. 3.25: 1) of the DT, the test field MMIC contains test structures to measure the transmission coefficients between the single-ended port on one and all differential ports on the other side of the DT (Fig. 3.25: 2 - 5).
Furthermore, there are two more test structures on the test field. The sixth breakout circuit is a realization of a 12.5 Ω line based on the LC (telegrapher) line principle [PDK$^+$14], consisting of short elements of air bridges (L) and MIM capacitors (C) which model an artificial line with a characteristic impedance of 12.5 Ω. Very short lines like this are used in each DT breakout circuit to connect the ports on the right side of the DT to their RF-pads. In a complete MMIC circuit, compact 12.5 Ω lines like this can be used to connect the DT to other circuit elements as will be seen in the next chapter. More common types of lines (CPW line, Microstrip (MS) line, etc.) are usually not able to support characteristic impedances that low without occupying a lot of wafer space due to their width. Therefore, at least for the applied processes, the LC line is one of the few possible ways of designing a compact line with a low characteristic impedance.
The last breakout circuit is a single 12.5 Ω resistor, which was implemented as a process monitor, to make sure that the DT ports, which are not connected to an RF-pad, are terminated correctly.
In Fig. 3.26 the measurement and the simulation results of the back-to-back configuration (Fig. 3.25: 1) of the DT are compared. A good agreement, especially for the frequency range between 0 and 170 GHz can be seen in all parameters. The H-band measurements show a deviation from the simulation, which is due to inaccuracies of the measurement that are unavoidable at such high frequen-

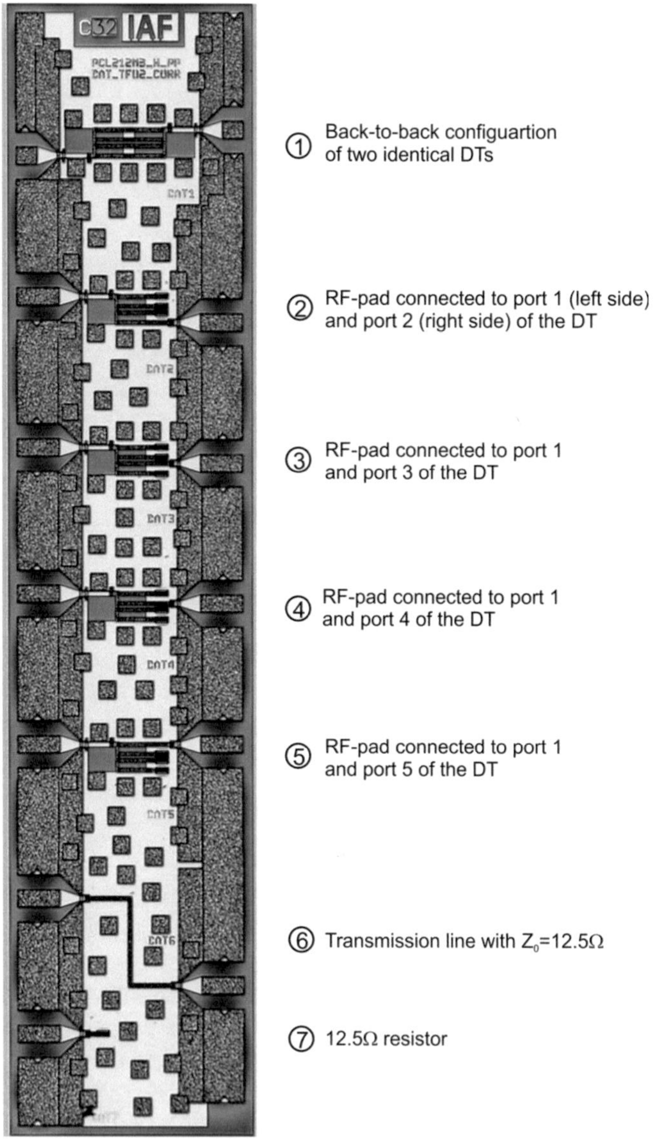

Figure 3.25: The measured test structure of a 5 port DT with a center frequency of 240 GHz.

cies. The relatively high fluctuation margin especially at the end of the H-band measurements also gives rise to this assumption.
The back-to-back configuration shows a minimal IL of 1.9 dB at 170 GHz. It has a relative 3 dB IL bandwidth of 85% (130 GHz - 325 GHz) and a relative 10 dB RL bandwidth of 71% (155 GHz - 325 GHz). This means that the actual single DT has only a minimum IL of 0.95 dB and an even slightly higher bandwidth. At 240 GHz the IL of a single DT is approximately 1.5 dB.
Based on the DT breakout circuits two to five, the amplitude and phase imbalance between port 1 and the four output ports (2-5) was measured as well. In Fig. 3.27 the transmission coefficients between port 1 and all output ports are plotted.
At 160 GHz and 240 GHz the various DT breakout circuits show the following transmission coefficients:

	EM in dB	$\angle$(EM) in $^\circ$	Meas. in dB	$\angle$(Meas.) in $^\circ$
S_{21} @ 160 GHz	-7.5	79.9	-7.8	56.4
S_{31} @ 160 GHz	-7.9	-95.3	-7.7	-117.2
S_{41} @ 160 GHz	-8.5	91.0	-8.7	69.5
S_{51} @ 160 GHz	-8.3	-91.7	-7.3	-103.6
S_{21} @ 240 GHz	-9.4	-1.7	-8.9	-11.2
S_{31} @ 240 GHz	-9.4	-171.7	-9.7	-173.2
S_{41} @ 240 GHz	-9.5	12.8	-9.9	0.6
S_{51} @ 240 GHz	-9.7	-171.8	-14.4	-174.6

Table 3.10: Simulation (EM) and measurement results of the DT breakouts for 160 GHz and 240 GHz.

As expected, the general shape of the transmission from port 1 to the four output ports (2-5) resembles that of the back-to-back configuration. Due to additional losses in the breakout circuits for the single ports, the simulated and measured loss is slightly higher than what was simulated and measured for a single DT of the back-to-back structure.
By comparing the various transmission coefficients listed in table 3.10 and depicted in Fig. 3.27, it is obvious that the measurement of test structure 5 does not behave

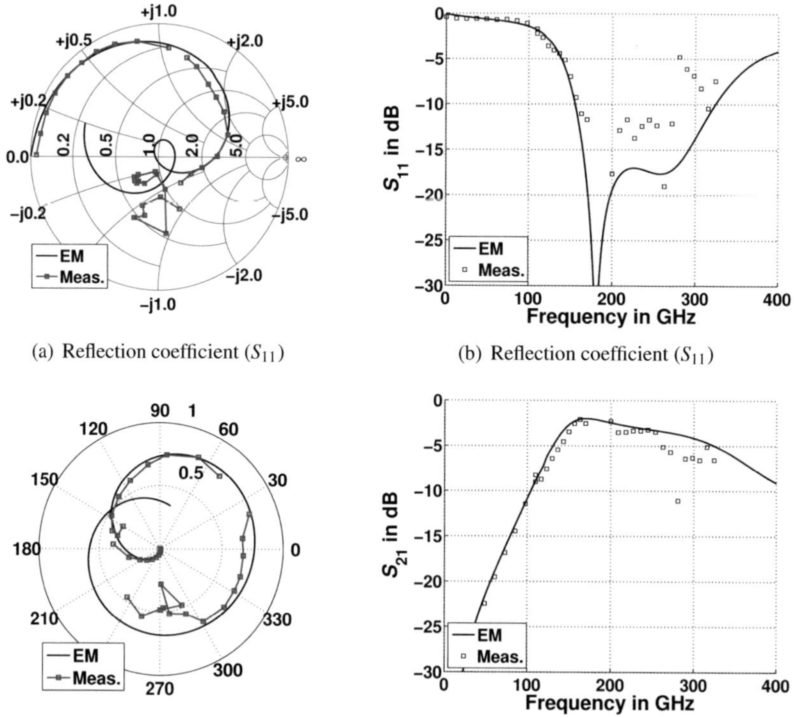

(a) Reflection coefficient (S_{11})
(b) Reflection coefficient (S_{11})
(c) Transmission coefficient (S_{21})
(d) Transmission coefficient (S_{21})

Figure 3.26: Simulation and measurement of a back-to-back structure of a 5 port DT with a center frequency of 240 GHz.

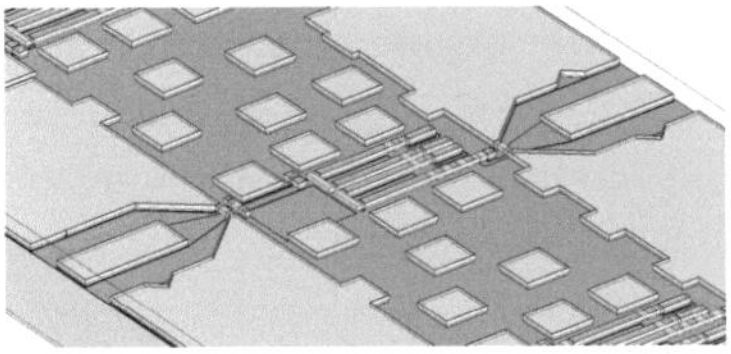

(a) Test structure port 2

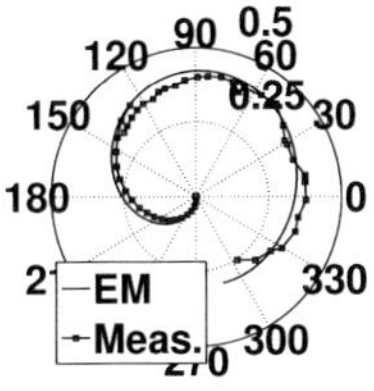

(b) S_{21}

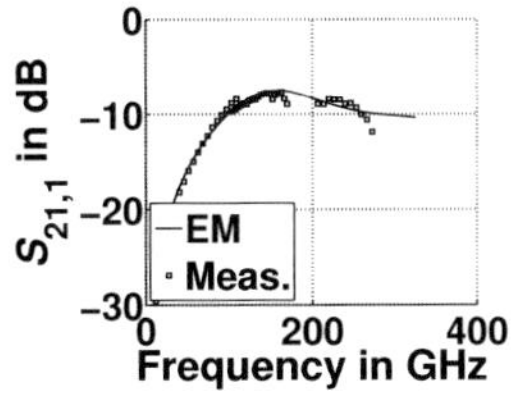

(c) S_{21}

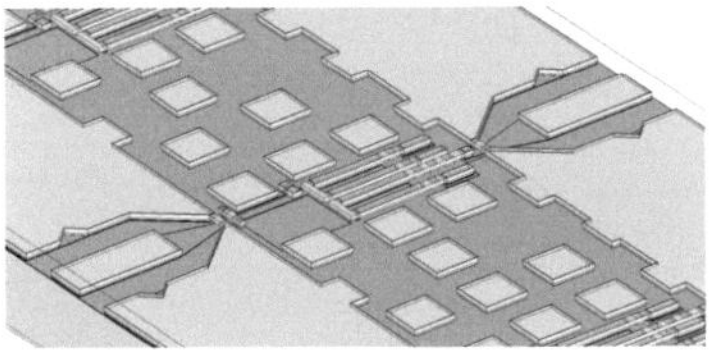

(d) Test structure port 3

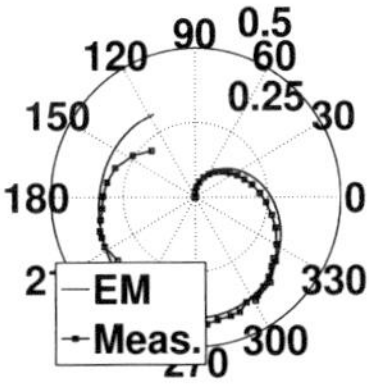

(e) S_{31}

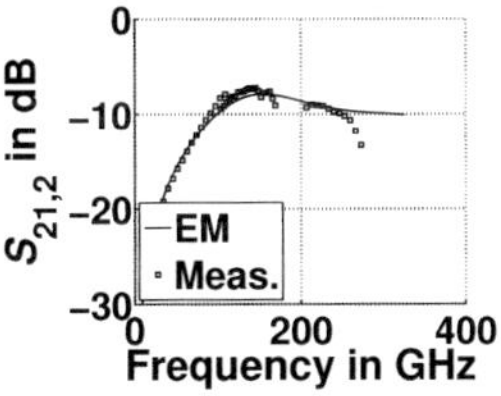

(f) S_{31}

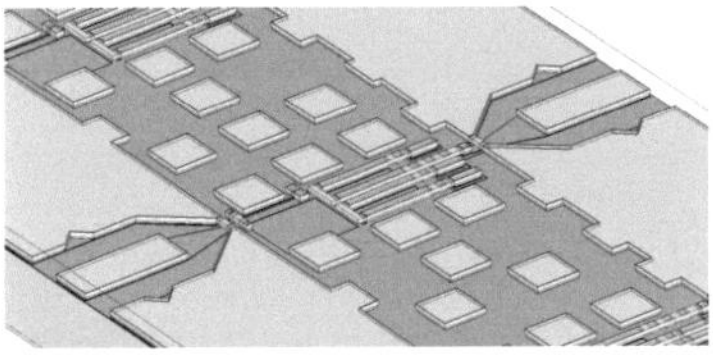

(g) Test structure port 4

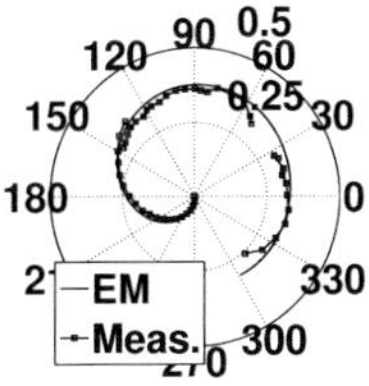

(h) S_{41}

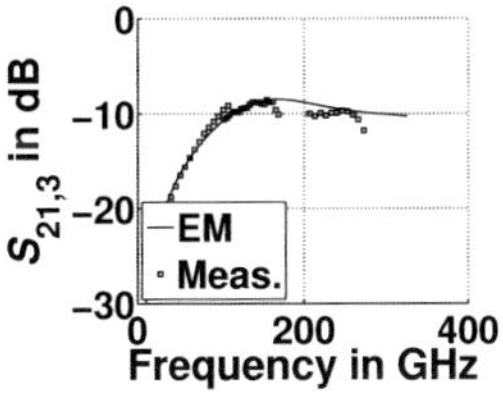

(i) S_{41}

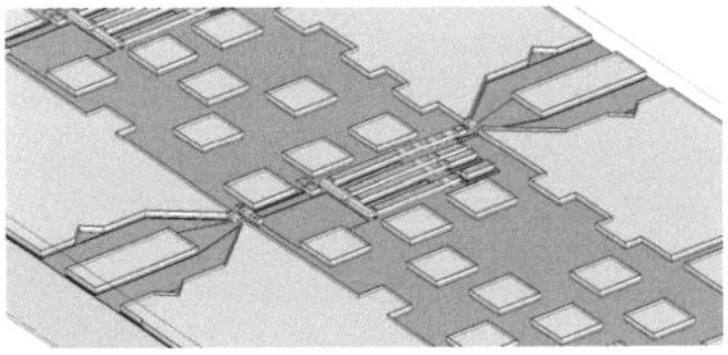

(j) Test structure port 5

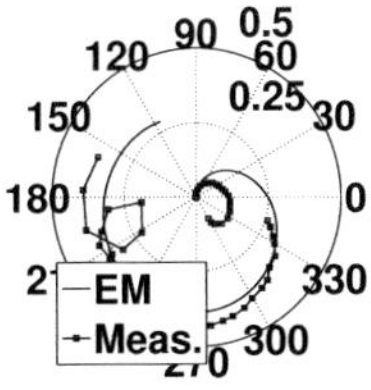

(k) S_{51}

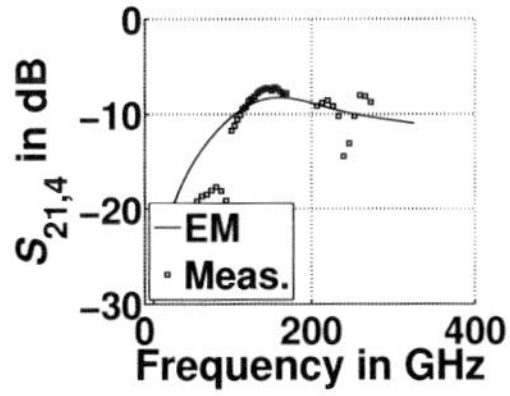

(l) S_{51}

Figure 3.27: Simulated and measured transmission coefficient (b,e,h,k) and insertion loss (c,f,i,l) for various test structures of 5 identical 5 port DTs with a center frequency of 240 GHz.

like its simulation. The reason for this is a layout error in the fabricated test field. For test structure 5 the load (12.5 Ω resistor) at port 2 was not connected, which leads to the visible deviation between measurement and simulation.
As final parameters of interest for the comparison between measured and simulated DT, the maximum amplitude and phase imbalance between the various ports can be compared. Due to the layout error, the measurement of test structure 5, does not give reasonable results and thus it was not considered for the measurement evaluation but only for the EM simulation. In Fig. 3.28 both, the maximum amplitude and phase imbalance of the ports of the DT are plotted. Measurement and simulation show an acceptable agreement.

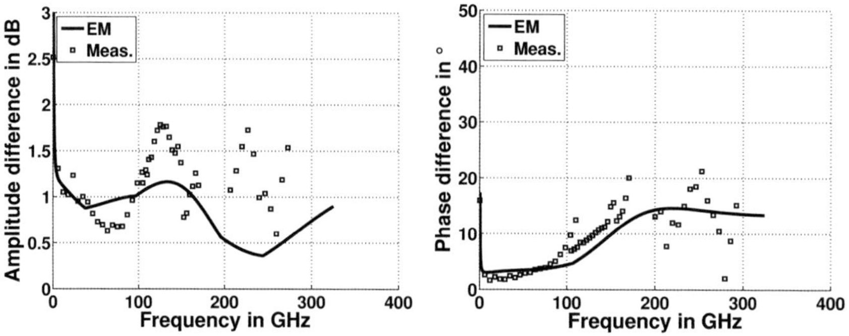

(a) Maximum amplitude difference in dB between the ports 2 to 5 for a signal injected at port 1.

(b) Maximum phase difference in degree between the ports 2 to 5 for a signal injected at port 1.

Figure 3.28: Comparison between the maximum amplitude and phase difference of a measured and EM simulated 5 port DT.

3.7 Conclusion on chapter 3

In chapter 3 the design and implementation of DTs based on the optimization equations of chapter 2 were demonstrated for Fraunhofer IAF's 35 nm and 50 nm mHEMT processes. The main achievement of the chapter is the presentation of a complete design flow to obtain an optimized DT.

The first step is the selection of the best type of coupled line in terms of: coupling coefficient (k_{m}), characteristic impedances ($Z_{\mathrm{e,o,c,}\pi}$) and quality factors (Q). A high k_{m} results in a wide symmetrical bandwidth and high efficiency. The efficiency also benefits from a high Q. To guarantee the amplitude and phase balance between the DT ports, a high characteristic impedance (i.e. Z_{e}, Z_{o}, Z_{c} and Z_{π}), i.e. a short DT is desired as well.
In a symmetrical coupled line the mean value of the characteristic impedance of both modes ($\frac{Z_{\mathrm{e}}+Z_{\mathrm{o}}}{2}$) acts like a scaling factor for the inductance per length, which means that a higher mean value leads to a shorter line in the optimization and therefore less amplitude and phase imbalance in the DT. For the asymmetrical coupled line, which has four characteristic impedances (one per propagation mode and independent line), the scaling factor is not just the mean value of all four impedances but a weighted mean ($\overline{\frac{Z_{\mathrm{c}}+Z_{\pi}}{2}}$), which to the author's best knowledge was first derived in this chapter. To evaluate the different coupled line types regarding their suitability for short DT unit cells and therefore a good amplitude and phase balance of the DT, the unweighted and weighted mean values for the characteristic impedances of the symmetrical and asymmetrical coupled lines of the different coupled line types are compared.

Three possible coupled line geometries were adopted from literature and implemented in the applied MMIC process. EM simulations were conducted for all three types of coupled lines and all of their possible geometry variations (i.e. various line widths (*cwidth*) and ground separations (*gnd_sep*) values). The obtained S-parameters were converted into per unit length quantities ($\mathrm{R}', \mathrm{L}', \mathrm{G}', \mathrm{C}'$), which were used to calculate and compare the DT parameters mentioned above. This analysis has shown that the asymmetrical coupled line geometry achieves the highest values in terms of coupling coefficient (k_{m}), quality factor Q and mean characteristic impedance ($\frac{Z_{\mathrm{e}}+Z_{\mathrm{o}}}{2}$, $\overline{\frac{Z_{\mathrm{c}}+Z_{\pi}}{2}}$). Thus, the asymmetrical coupled line type was chosen for all DT unit cell designs in this thesis.

A DT consists of multiple DT unit cells, which can be connected in various ways. Whether DT unit cells are put in parallel or in series to each other and which ports of the DT unit cells are connected to one another or serve as ports of the DT determines the impedance transformation ratio of the DT and also the phase relation between the different DT ports. Various configurations have been introduced and demonstrate that the DT concept can be used to build an impedance transforming power dividing multiport balun, that achieves a wide number of impedance transformation ratios and phase combinations for a given number of ports.

For the objective of an impedance transforming power divider/combiner as part of a PA, the voltage combiner type, which puts all DT unit cells in series on its single-ended side and in parallel on the multi-port, i.e. differential side is the most convenient one, as it provides the highest impedance transformation ratio for a given number of input/output ports. The voltage combiner type is also the prevailing type of DT/DAT in literature.

Since all published DT/DAT based PAs are built on Si based technologies the single DT unit cells are usually arranged in a circle to make use of the virtual grounds between adjacent transistors (push-pull configuration). However, by using a III-V compound semiconductor technology the issue of lossy GND connections is mitigated, which enables a novel straight DT configuration presented in this chapter. The new DT geometry affects the behavior of the single optimized DT unit cells applied in the DT less than the circular geometry. Therefore, it is more suitable for the design procedure presented in this thesis, which assumes that the arrangement of the unit cells in the DT does not alter the frequency response of the unit cells. The new geometry also supports the distribution of the signals better than the circular geometry does, since the ports on the multi-port side of the DT are further apart from each other and can all be on the same side of the DT.

As an example for a possible implementation, a complete 1:4 DT layout was proposed in the chapter. It is based on asymmetric coupled line unit cells and implements a voltage combiner type DT in form of the novel straight DT geometry. The applicability of the design equations for maximum efficiency and maximum bandwidth has been verified on basis of the proposed DT layout. The design procedure presented in this chapter assumes that the frequency response of a DT is mainly determined by the DT unit cells. Based on the per unit-length quantities for all possible asymmetrical coupled line configurations (various line widths and ground separations), which have been obtained from EM-simulations, and the optimization equations of chapter 2, the optimized behavior of the complete DT can be predicted for every configuration without simulating the whole DT. By evaluating the DT optimization results, an acceptable trade-off between efficiency and bandwidth as well as amplitude and phase balance of the DT can be chosen in form of a DT unit cell geometry and implemented for instance in the DT layout proposed in this chapter. The applicability of the novel design flow was proven by the comparison of the behavior of DTs determined only from EM simulations of a single coupled line segment and that of the EM simulation of the entire DT layout for multiple DTs optimized for maximum efficiency and maximally flat IL BW. All of these comparisons showed a very good agreement.

The last part of the chapter presented a comparison between EM simulation and measurement of a 5 port DT. It discussed the difficulties that arise by measuring DTs at high mmW frequencies and demonstrated that for an accurate measurement of the transmission coefficient between the single-ended port on one and one of the multiport/differential ports on the other side of the DT, a de-embedding of the RF-pads is inevitable. By comparing the Open-Short method de-embedding with the de-embedding based on EM simulated RF-pads and their ABCD-parameters, it was found that the Open-Short method cannot be used for a de-embedding above 110 GHz, while the ABCD-parameter method works properly over the entire frequency range. By applying the ABCD-parameter de-embedding method a comparison between the EM simulation and the measurement of a DT based IMN concluded the chapter. The demonstrated 1:4 DT (5 port) uses the straight DT layout proposed in this chapter in combination with asymmetrical coupled line unit cells to demonstrate the wide relative bandwidth and high efficiency that can be achieved for IMNs/OMNs based on DTs.

Chapter 2 and 3 together provide the means to design highly efficient and very broadband IMNs and OMNs that can be applied beneficially in various RF circuits. The focus of this thesis is their application in IMNs and OMNs of broadband PAs, which will be the topic of the next chapter.

4 Power Amplifiers Based on DTs

In the previous chapters, design strategies for DTs have been introduced. Based on the theoretical considerations of chapter 2, which are valid for all types of transformer based IMN and OMN, chapter 3 proposed a DT layout, optimized for MMICs operating at mmW frequencies on two top-metal-layer processes. In this chapter, the design of power amplifiers based on DT will be discussed and demonstrated. Based on two different broadband gain cell typologies, two state-of-the-art PAs will be presented, that exhibit the broadest bandwidth ever reported for power amplifiers at these frequencies to date and benefit greatly from the features of the DT.

4.1 Concepts for power amplifiers based on DTs

In the introduction (chapter 1) it was mentioned, that power amplifiers usually apply multiple transistors in parallel to overcome the power limitation of a single transistor device. Furthermore, transistors operated at high millimeter-wave frequencies show low input and output impedances to which the relatively high PA port impedance (in general 50 Ω) is matched by an appropriate matching network. The DT presented in the previous chapters can be applied to do both of these things simultaneously, by transforming one high into multiple low impedance ports. Depending on the DT configuration these ports can be differential, which can be advantageous for the design of the PA's gain cells.
In most of the PA designs published in literature, the DT is mainly applied as power combiner and matching network in the last stage of the PA. There, the parasitic inductance of the transformer is usually used to compensate the capacitive transistor output impedance. In Fig. 4.1 the schematic of a three port (1:2) DAT

is depicted. To illustrate the matching concept, a simple ECM of the drain side of a FET is connected to each input port of the DT's ECM. C_p and L_p are chosen in a way that the parallel resonator consisting of $C_{ds} + C_p$ and L_p resonates at the operation frequency of the PA.

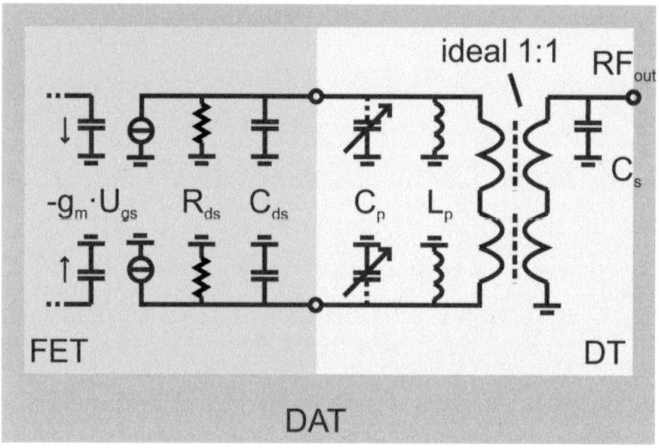

Figure 4.1: Principle ECM of a DAT based PA.

Due to the minimum number of passive matching elements, the DAT concept is very efficient. On the other hand, the minimum number limits the degree of freedom to compensate the frequency response of the transistor. As already mentioned in chapter 1, a low order OMN, e.g. the 2. order network of a DAT, limits the possibilities to achieve a wide bandwidth. Therefore, the DAT concept is not ideal for broadband power amplifiers, even though the DT in general provides a low impedance over a wide frequency range.

The DAT/DT concept cannot only be used to realize efficient OMNs, but is equally suitable for efficient IMNs as well. If DTs are applied for the design of power amplifiers on III-V semiconductor MMIC technologies, the most convenient configuration for a PA with for instance four parallel branches is the one depicted in Fig. 4.2, where identical DTs are used in the IMN and OMN to design a very compact PA.

Depending on the required bandwidth, the gain cells can be very simple. In the simplest case, the PA consists only of the two DTs at the in- and output and four

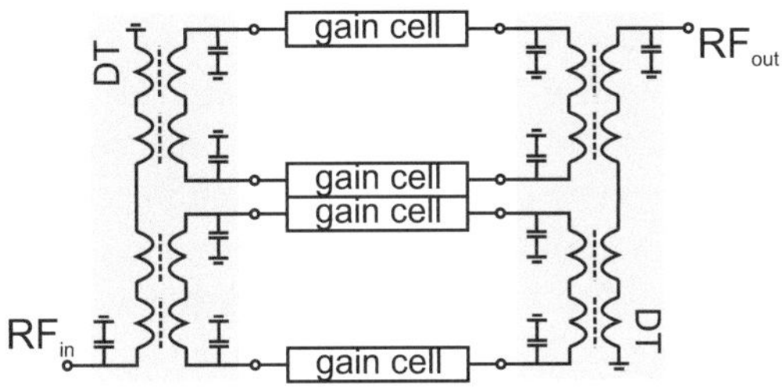

Figure 4.2: Proposed PA design based on two DTs and multiple gain cells.

parallel transistors in between, which means it consists of two DATs. However, one limiting factor of such a configuration at high millimeter-wave frequencies is the maximum stable/available gain of the applied transistors. Given the loss of the DT, most of the time it is not high enough to achieve the desired overall amplification.

By using not only one but multiple gain cells in series, in between the DTs, the overall gain and bandwidth of the PA can be increased. In this case, more complex but also more lossy gain cells with additional matching and bias networks are necessary. The easiest way to cascade multiple gain cells is to use identical gain cells with a fixed real reference impedance at each stage's in- and output port. Because the DT can provide a low port impedance over a wide frequency range, it eases the design of an efficient broadband matching network by reducing the number of necessary reactive elements for a broadband matching. Ideally, the reference impedance of the gain cell should be identical to the real part of the input or output impedance of the active device at the upper edge of the frequency range of interest. This way a minimum number of reactive elements would be sufficient to compensate the reduction of MSG/MAG over a very wide frequency range by altering the matching of the active elements respectively. The closer the gain cell's port impedance is to the actual required transistor input and output impedance, for a desired gain shape and matching, the less the actual port termination of the active devices deviates from the desired termination, for a given number of reactive matching elements in the gain cells matching networks, over frequency.

In other words, if the order of the gain cell's IMN and OMN is fixed, a wider bandwidth can be achieved, if the reference/port impedance of the gain cell is closer to the actual desired termination of the active elements within the gain cell. Therefore, the DT's ability to provide a broadband impedance transformation enables the design of more broadband PAs by simultaneously providing the required parallelization. That the DT can be operated as a balun adds another feature, which can be beneficial, as will be seen in the sections ahead.

4.2 Measurement results of complete PAs

In the next two subsections, the design of two PAs based on DTs will be discussed. The amplifiers operate between 200 GHz and 270 GHz and 50 GHz and 146 GHz, respectively. Both amplifiers apply three port DTs, that transform the 50 Ω reference impedance at the ports of the PAs into 25 Ω at the ports of the utilized gain cells. The first design uses multiple standard LC matched gain cells while the second applies a traveling wave amplifier (TWA) gain cell in each parallel branch of the PA to achieve the desired bandwidth.

4.2.1 PA based on a low impedance LC amplifier gain cell

In Fig. 4.3(a) the principle schematic of an R-band (170 GHz - 260 GHz) amplifier is depicted. It consists of two identical DTs, one at the in- and one at the output of the PA and three identical gain cells cascaded in between. Due to the chosen configuration of the DT unit cells, the DT transforms the input signal of a 50 Ω single-ended port into two differential signals at two 25 Ω ports. By means of two short 25 Ω lines, the out-of-phase signals are connected to the first of three consecutive pseudo-differential gain cells. The detailed schematic of the gain cells is depicted in Fig. 4.3(b).
On the input and output side of the gain cell the matching is achieved by parallel shorted stubs and series lines as shown in the schematic. The parallel stubs are not only applied for matching purposes but also provide an elegant way of supplying each of the two branches with the required direct current (DC) power. Due to the

differential signals on both branches a direct connection of their parallel stubs results in a virtual ground at their intersection, which isolates each of the two branches from each other, even though they are directly connected. By avoiding the necessity of supplying each of the two branches individually, the overall layout of the gain cell can be very compact. This advantage would become even more significant, if there were more than two parallel branches, which means that there are branches in the middle of the gain cell, whose DC bias connection would be complicated otherwise. Therefore, the differential signals provided by the DT can be very useful for the design of compact gain cells.
As active element of each gain cell, a cascode consisting of one transistor in CS configuration and a second in common-gate (CG) configuration is implemented. Both transistors have a gate length of 50 nm (IAF's 50 nm MMIC technology), two fingers and a gate width of 10 µm per finger resulting in a total gate width of 20 µm per transistor. Since the gain of one cascode stage is not sufficient to provide an adequate amount of gain, three gain cells are cascaded.
After the third and last stage, two short 25 Ω lines connect the gain cell with a second DT, which is identical to the one at the input of the PA. The DT at the output recombines the signals of both parallel branches and provides the 25 Ω termination for each branch's last gain cell, if the PA is terminated with 50 Ω.
To verify the performance of the various parts of the DT based PA individually, multiple breakout structures of the PA were fabricated and measured. In Fig. 4.4(a) and 4.4(b) two photographs of the applied DT with RF-pads connected to port 1 and either port 2 or 3 of the DT are shown. These test structures were used to evaluate the IL of the DT and the power distribution between the two ports of the DT. The simulated and measured S-parameters are shown in Fig. 4.4(c).
As visible from the S-parameters, the DT is very broadband and shows a reasonable loss over its entire bandwidth. At 220 GHz the breakout structures for the assessment of port 2 and 3 have a minimum insertion loss of 3.8 dB and 3.6 dB, respectively. From 130 GHz up to more than 400 GHz the insertion loss never increases above 6.6 dB, which corresponds to a 3 dB bandwidth of more than 270 GHz, i.e. a relative bandwidth of more than 100 %.
In Fig. 4.5 the breakout circuit of the gain cell is investigated. A photograph of the measured circuit is shown in Fig. 4.5(a) and the measured and simulated S-parameters are compared in Fig. 4.5(b) for the three cascaded gain cells of one

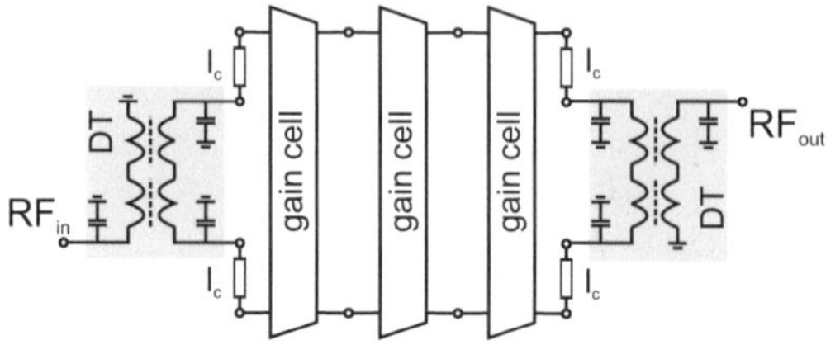

(a) Schematic of the amplifier.

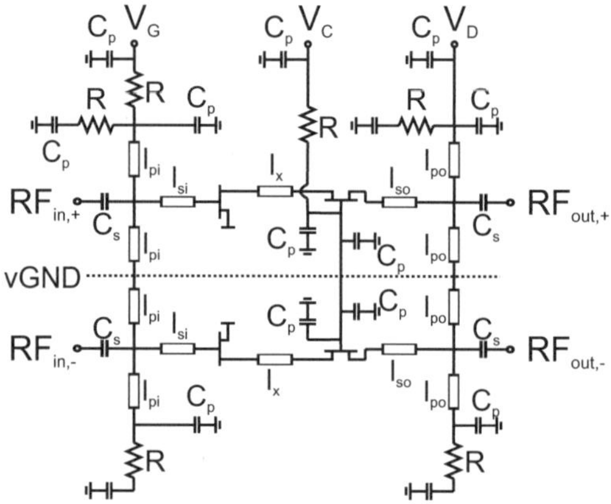

(b) Schematic of the gain cell.

Figure 4.3: Schematic of the PA based on low impedance LC gain cells.

branch of the PA and in Fig. 4.5(c) for a single gain cell. All S-parameters were measured at a drain current of $I_D = 400\,\mathrm{mA/mm}$ and a drain voltage of $V_D = 1.6\,V$. Both, the single gain cell as well as the three cascaded gain cells show good agreement between simulation and measurement. The ripple in the measurement of the three cascaded gain cells is attributable to unwanted feedback between the input and output ports of the test structure during the on-wafer measurements that usually disappear after dicing of the chip. By simulating a very small capacitance between the ports of the gain cell, the ripple can be reproduced in the simulation

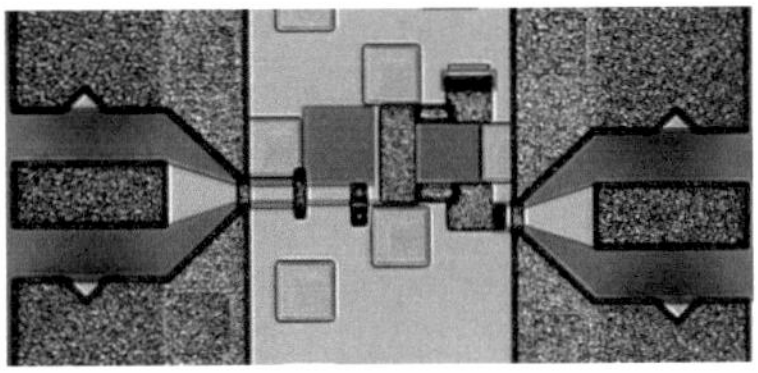

(a) Chip photograph of the DT breakout to test port 2.

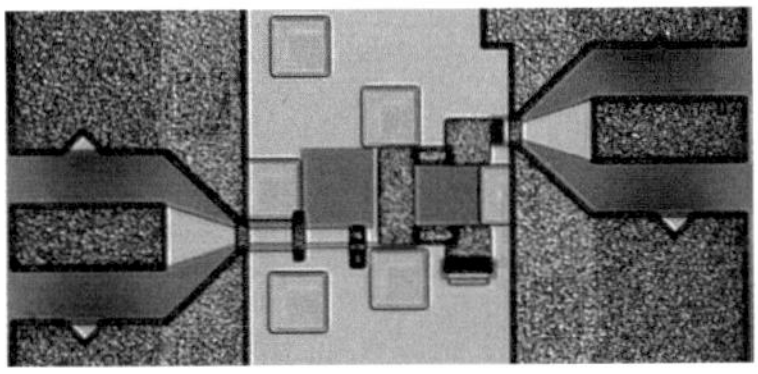

(b) Chip photograph of the DT breakout to test port 3.

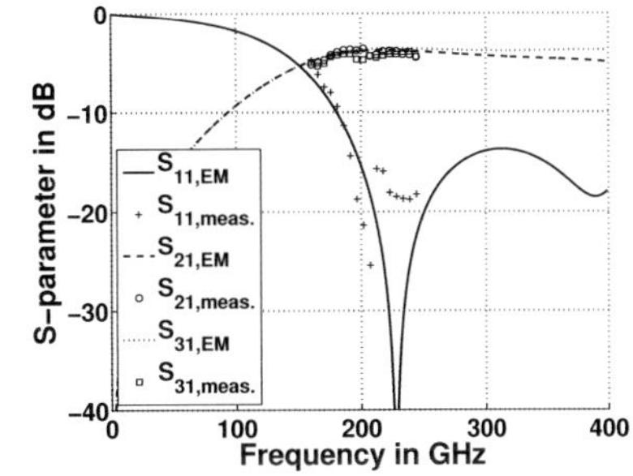

(c) Measured and simulated (EM) S-parameters of the DT breakout circuits to test port 2 & 3. ($Z_{0,p1} = 50\,\Omega$, $Z_{0,p2} = Z_{0,p3} = 25\,\Omega$)

Figure 4.4: DT evaluation for the PA based on low impedance LC gain cells.

as well. Due to the gain ripple, the -3 dB bandwidth of the measurement is not reliable, which is why the simulated gain and bandwidth for three cascaded gain cells will be listed here. The maximum gain is 28.8 dB and the -3 dB bandwidth extends from 190 GHz to 272 GHz. This equals a relative bandwidth of 30.5 %.

By comparing the passband of the DT with the frequency range of the gain cells, it becomes obvious that the limiting element regarding bandwidth of the PA is the gain cell. If a gain cell with a wider bandwidth would be used, the amplifier could have a wider bandwidth without modifying the DT. At least towards higher frequencies the frequency range of the PA could be easily doubled, with the same DT.

The photograph of the entire PA is shown in Fig. 4.7. It is built with the DT and the gain cells presented before and connected in the way shown in Fig. 4.3.

Biased at exactly the same quiescent point as the gain cells before ($I_D = 400\,\mathrm{mA/mm}$ and $V_D = 1.6V$), the PA demonstrates a gain of 29.2 dB at 237 GHz. Its 3 dB

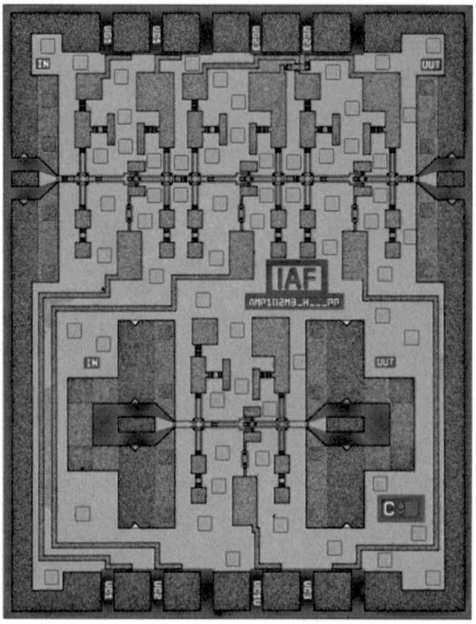

(a) Chip photograph of the gain cell test field MMIC. (size: 1 x 1.25 mm^2)

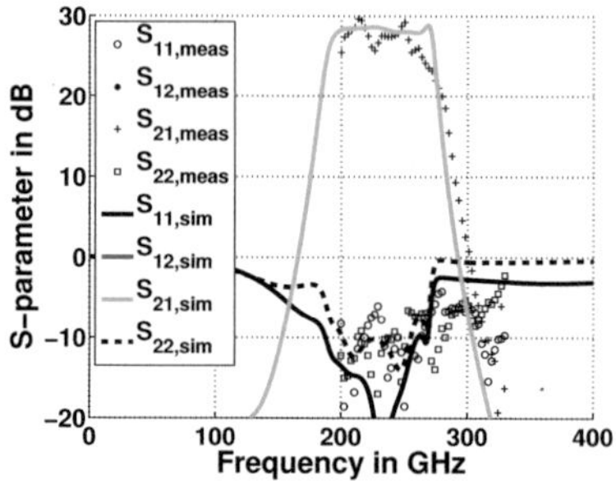

(b) Measured and simulated S-Parameters of the gain cell. ($Z_0 = 25\,\Omega$, $I_D = 400\,\text{mA/mm}$, $V_D = 1.6\,\text{V}$)

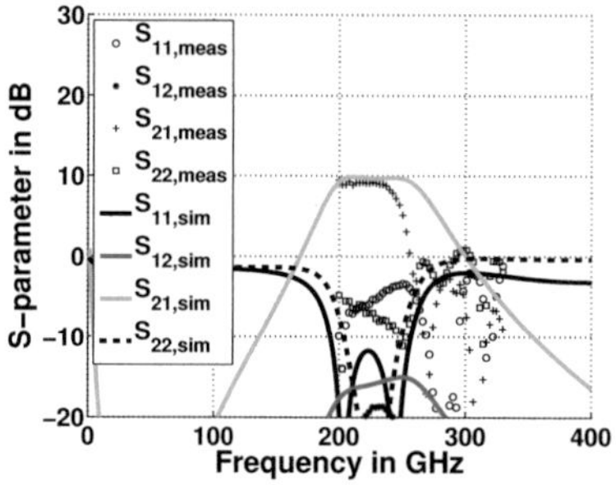

(c) Measured and simulated S-Parameters of one stage of the gain cell. ($Z_0 = 25\,\Omega$, $I_D = 400\,\text{mA/mm}$, $V_D = 1.6\,\text{V}$)

Figure 4.5: Gain cell evaluation of the PA based on low impedance LC gain cells.

bandwidth goes from 201 GHz - 270 GHz, which is equal to a relative bandwidth of 29 %. The visible gain ripple in the measurement is again due to the high gain and the relatively low isolation between the input and output ports at high mmW frequencies during on-wafer measurements, which lead to a strong effect of the unwanted feedback on the gain response of the PA. After dicing, the ripple is expected to disappear, but even with the ripple the PA already demonstrates the

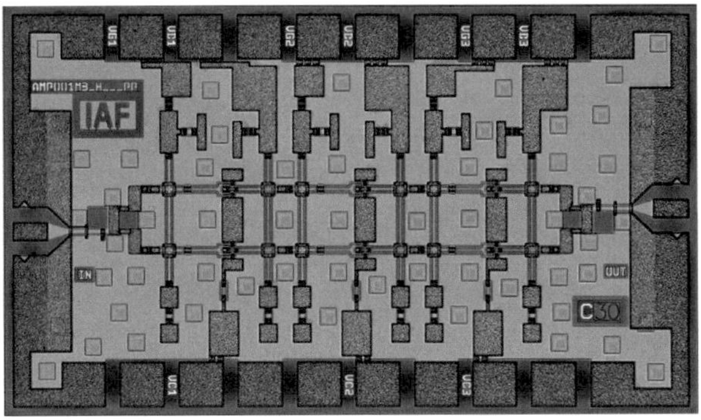

Figure 4.6: Chip photograph of the PA MMIC for the PA based on low impedance LC gain cells. (size: 1.25 x 0.75 mm^2)

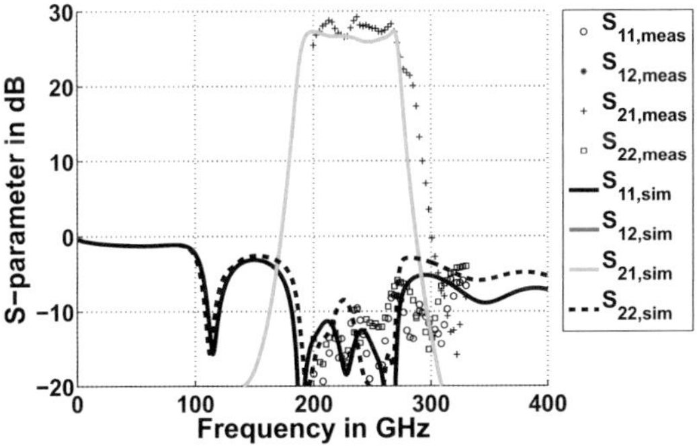

Figure 4.7: Measured and simulated S-parameters for the PA based on low impedance LC gain cells. ($Z_0 = 50\,\Omega$, $I_D = 400\,\text{mA/mm}$, $V_D = 1.6\,\text{V}$)

largest -3 dB bandwidth ever reported for a PA at this frequency range to date. Large signal measurements of the PA were also conducted. For these measurements, the PA was operated at a drain current of $I_D = 400\,\text{mA/mm}$ and a drain voltage of $V_D = 2.2V$. In Fig. 4.8 the output power (P_{out}) and the large signal gain are plotted vs. input power (P_{in}) at 260 GHz. The amplifier shows a maximum measured output power of 4.9 dBm with a large signal gain of 20 dB. For a PA in Fraunhofer IAF's 50 nm mHEMT technology with two parallel transistors in the output stage (total gate with of 40 µm), an output power of 4.9 dBm (power density of 77 $\frac{mW}{mm}$) is only sightly less than what has been reported in other PAs with the same technology [LDM+13] but unlike those amplifiers the presented PA has a much wider 3 dB bandwidth.

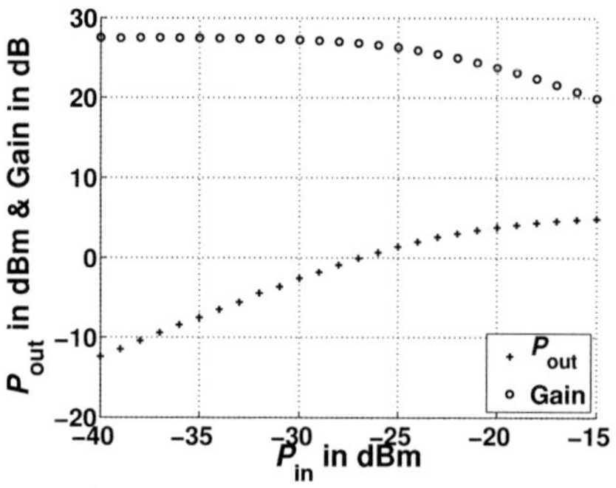

Figure 4.8: Measured large-signal parameters at 260 GHz for the PA based on low impedance LC gain cells. ($I_D = 400\,mA/mm$, $V_D = 2.2\,V$)

To get an idea about the frequency dependency of the output power, Fig. 4.9 plots the 1 dB compression point in- and output power (Fig. 4.9(b)) and the output power for a constant input power of -15 dBm, -20 dBm and -25 dBm (Fig. 4.9(b)) versus frequency between 200 GHz and 260 GHz.

The output related 1 dB compression point over the frequency range of interest is about 1 dBm. For a constant input power the output power varies within the range of 2 dB. The reason for the output power ripple is the unwanted feedback, due to on-wafer power measurements. Over the entire frequency range of interest the PA provides at least 3 dBm of maximum output power. It can be observed that the maximum output power increases towards the upper edge of the amplifier's passband.

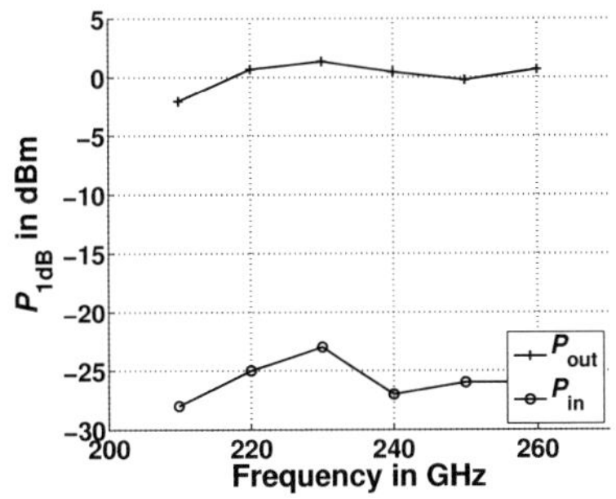

(a) Measured 1 dB compression point versus frequency from 210 GHz to 260 GHz. ($I_D = 400\,\mathrm{mA/mm}$, $V_D = 2.2\,\mathrm{V}$)

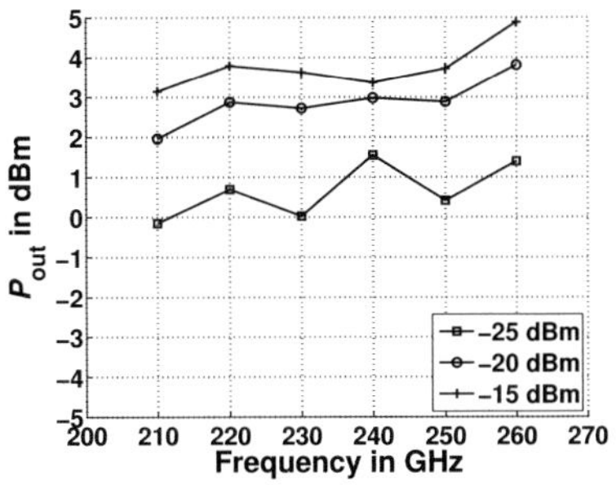

(b) Measured output power versus frequency from 210 GHz to 260 GHz. ($I_D = 400\,\mathrm{mA/mm}$, $V_D = 2.2\,\mathrm{V}$)

Figure 4.9: Output power vs. frequency for the PA based on low impedance LC gain cells.

4.2.2 Traveling wave amplifier based PA

Regarding the bandwidth, the PA presented in the previous subsection was primarily limited by the bandwidth of its gain cell. As mentioned above, the DT would have supported a much wider bandwidth. To demonstrate a PA, which really takes advantage of the wide bandwidth of the DT, a PA based on TWA gain cells will be presented next. The schematic of the PA is depicted in Fig. 4.10.

The PA consists of two TWA gain cells in parallel, which deploy seven TWA unit cells each. The TWA unit cells use airMS lines as inductive lines for the TWA's artificial gate and drain lines. As active elements cascode stages are used, which consist of one CS and one CG mHEMT. The CS and CG transistors both have a gate length of 35 nm, two fingers and a gate width of 20 µm per finger. By not feeding the TWA gain cells through the resistive terminations of the artificial transmission lines but through a vGND point of the DT, the overall PA's efficiency is improved compared to standard TWAs. On the gate side of both TWA gain cells, parallel capacitors provide the necessary RF GND and DC open, which

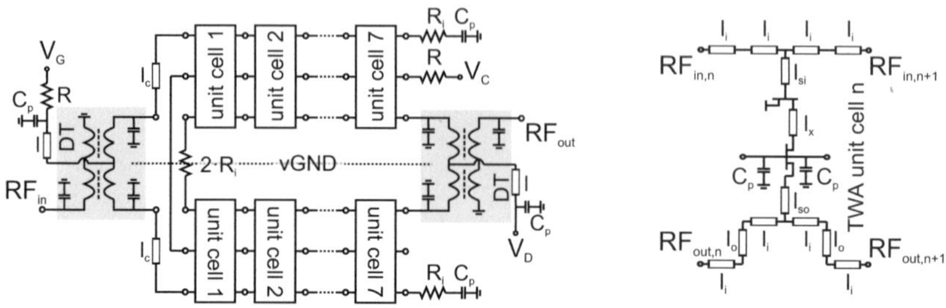

(a) Schematic of the amplifier.

(b) Schematic of the TWA unit cell.

Figure 4.10: Schematic of the low impedance TWA based amplifier.

together with a resistor terminate the artificial transmission line properly. The resistors are not grounded directly, because this would cause unwanted DC losses. On the drain side the required termination of the artificial line is realized by taking advantage of the vGND between gain cells again.

As for the low impedance amplifier presented in the previous section, the applied DT transforms the 50 Ω reference impedance at the input of the PA into two out of phase 25 Ω ports. After two parallel TWA gain cells, an identical DT recombines the amplified signals and provides two 25 Ω port terminations for the gain cells, if the PA is terminated with 50 Ω. Therefore, the gain cells of the PA operate in a 25 Ω environment, which, as for the low impedance amplifier in the previous subsection, enables a wider bandwidth compared to TWA gain cells operated at 50 Ω.

Fig. 4.11 depicts a simple ECM of a TWA. The gate and drain lines are represented by the inductors L_g and L_d. Instead of the FETs, their gate-source capacitance (C_{gs}), drain-source capacitance (C_{ds}) and current source are included in the schematic. The inductive lines in combination with the internal capacitors of the FETs build an artificial transmission line at the in- and output of the TWA. Since the single elements of the artificial lines are finite, i.e. lumped and not distributed, the artificial transmission line has a cut-off frequency and resembles a cascadation of N low-pass circuits.

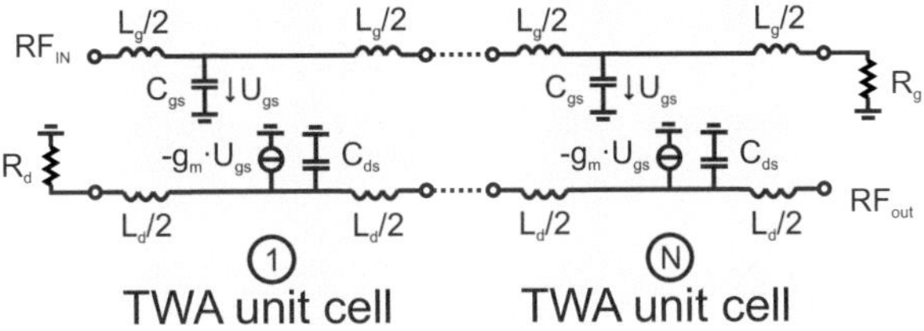

Figure 4.11: Simple schematic of a FET based TWA.

Using the image parameter method [Poz04], which is usually applied in filter theory, the cut-off frequency as well as the characteristic impedance of the artificial transmission lines can be determined [Ell05, VVB04]:

$$f_{\mathrm{c,g}} = \frac{1}{\pi\sqrt{L_{\mathrm{g}}C_{\mathrm{gs}}}} \quad f_{\mathrm{c,d}} = \frac{1}{\pi\sqrt{L_{\mathrm{d}}C_{\mathrm{ds}}}} \tag{4.1}$$

$$Z_{0,\mathrm{g}} = \sqrt{\frac{L_{\mathrm{g}}}{C_{\mathrm{gs}}}} \quad Z_{0,\mathrm{d}} = \sqrt{\frac{L_{\mathrm{d}}}{C_{\mathrm{ds}}}} \tag{4.2}$$

By combining Eqn. 4.1 and 4.2, it becomes clear that for a given transistor size and therefore a given C_{gs} and C_{ds}, the cut-off frequency depends on the characteristic impedance of the artificial transmission line.

$$f_{\mathrm{c,g}} = \frac{1}{\pi Z_{0,\mathrm{g}}C_{\mathrm{gs}}} \quad f_{\mathrm{c,d}} = \frac{1}{\pi Z_{0,\mathrm{d}}C_{\mathrm{ds}}} \tag{4.3}$$

The lower the characteristic impedance of the artificial transmission line, the higher the cut-off frequency and therefore the bandwidth of the TWA. By reducing the artificial characteristic impedance further and further the TWA's upper frequency limit is eventually determined by the cut-off frequency (f_t) of the applied semiconductor process.

In Fig. 4.12 a photograph of the TWA gain cell and its measured and simulated S-parameters are shown. The gain cell has an average small-signal gain of 13.5 dB over a bandwidth of 149 GHz (10 GHz - 159 GHz) in the simulation. This equals a relative bandwidth of 176 %. To compensate the negative gain slope of the two DTs before and after the gain cell of the PA, the gain cell's gain slightly increases towards higher frequencies. For the S-parameter measurements a quiescent point of $I_D = 500\,\text{mA/mm}, V_D = 1.4\,\text{V}$ was selected.

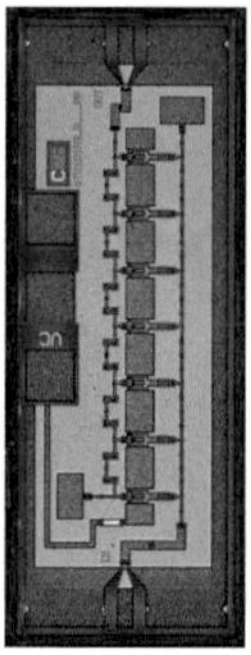

(a) Chip photograph of the gain cell of the traveling wave amplifier. (size: 0.5 x 1.25 mm^2)

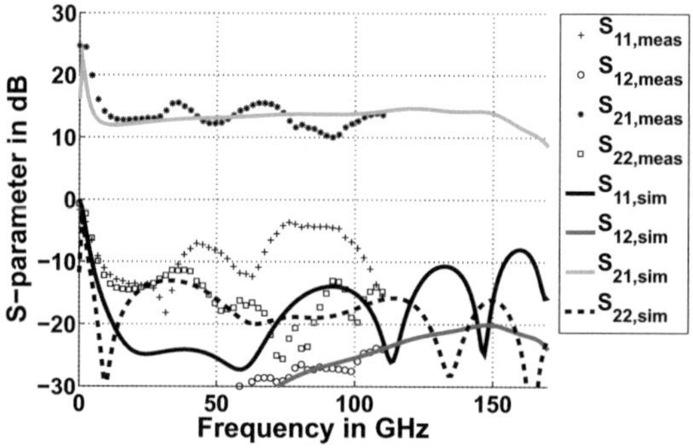

(b) Measured and simulated S-parameters of the gain cell of the traveling wave amplifier. ($Z_0 = 25\Omega$, $I_D = 500\,\text{mA/mm}$, $V_D = 1.4\,\text{V}$)

Figure 4.12: Gain cell of the traveling wave amplifier.

The DT presented in Fig. 4.13 is applied as power divider/combiner and impedance transformer in the PA. In the photograph of the back-to-back breakout circuit (Fig. 4.13(a)) of the DT, the DC supply lines, connected to the vGND point of the DT, can be seen. Their influence on the performance of the DT is negligible. The DT demonstrates a high bandwidth and an acceptable loss, visible in Fig. 4.13(b). The measured and simulated S-parameters of the DT show a relative bandwidth of 100 % (56.5 GHz - 170 GHz) and 94 % (56.4 GHz - 156.8 GHz), respectively. Within these frequency ranges the average loss of the back-to-back structure is 3.6 dB (1.8 dB per DT) in the measurement and 2.6 dB (1.3 dB per DT) in the simulation. Around 80 GHz the DT shows its highest efficiency of -2.5 dB (@ 82 GHz) in the measurement and -1.6 dB (@ 78 GHz) in the simulation of the back-to-back structure. Towards higher frequencies the DT has the typical

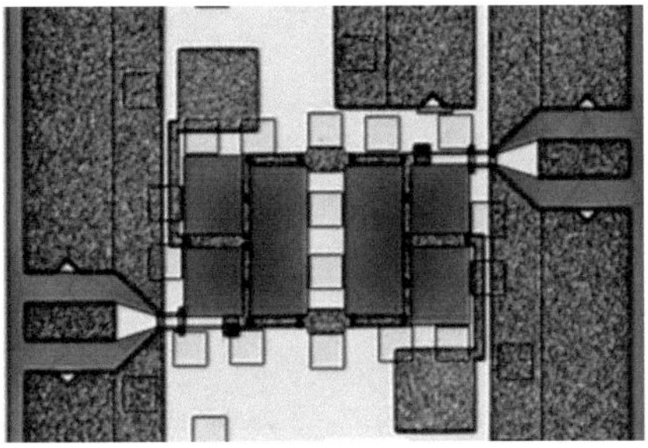

(a) Chip photograph of the DT in back-to-back configuration used in the traveling wave amplifier.

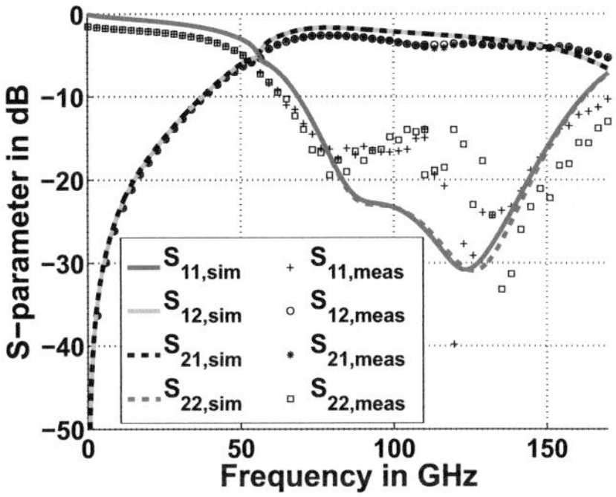

(b) Measured and simulated S-Parameters of the DT in back-to-back configuration used in the traveling wave amplifier. ($Z_0 = 50\Omega$)

Figure 4.13: DT of the traveling wave amplifier.

negative gain slope, which in combination with the positive gain slope of the gain cell results in a flat response for the complete PA.

A photograph of the entire PA circuit, including the two DTs, two gain cells in parallel, additional lines and RF/DC pads, is presented in Fig. 4.14. The RF signal is applied at the left and leaves the circuit at the right side of the MMIC. The DC power is supplied by DC pads at the top and the bottom of the chip.

In Fig. 4.15 the simulated and measured S-parameters of the PA are shown.

The PA has a maximum measured small-signal gain of 12 dB at 106 GHz. Its 3 dB bandwidth goes from 50 GHz to 146.65 GHz (96.65 GHz), which corresponds to a relative bandwidth of 98.3 %. Within this bandwidth the average gain is 10.8 dB. The quiescent point for the S-parameter measurement of this PA was $I_D = 500\,\text{mA/mm}$ and $V_D = 1.8\,\text{V}$.

Besides small-signal measurements, large-signal measurements have been conducted over the entire bandwidth of the PA. As an example Fig. 4.16 demonstrates the output power (P_{out}), PAE and large-signal gain versus input power (P_{in}) at 90 GHz and 140 GHz. The quiescent point for the large-signal measurements

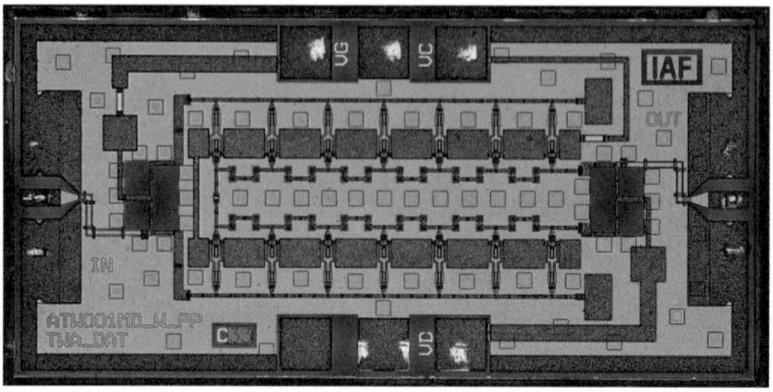

Figure 4.14: Chip photograph of the complete traveling wave amplifier. (size: 1.5 x 0.75 mm^2)

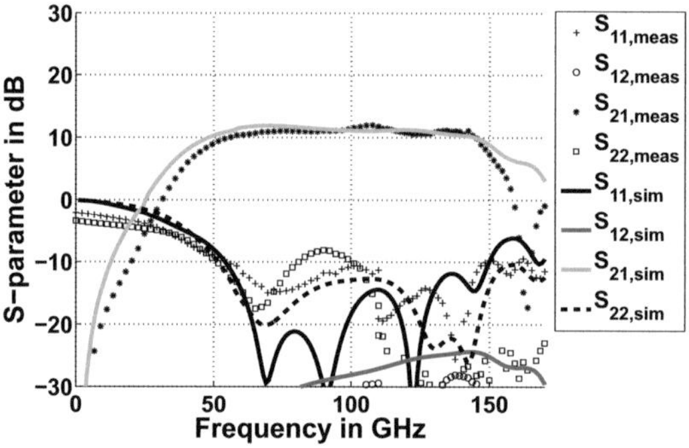

Figure 4.15: Measured and simulated S-parameters of the complete traveling wave amplifier. ($I_D = 500\,\mathrm{mA/mm}$, $V_D = 1.8\,\mathrm{V}$)

equals the bias point for the small-signal measurements: $I_D = 500\,\mathrm{mA/mm}$ and $V_D = 1.8\,\mathrm{V}$.

To give an idea about the variation of the output power versus frequency, Fig. 4.17 visualizes P_{out} versus frequency. In Fig. 4.17(a) the output power for an input power of 8 dBm is plotted in 1 GHz steps from 80 GHz to 110 GHz. Above

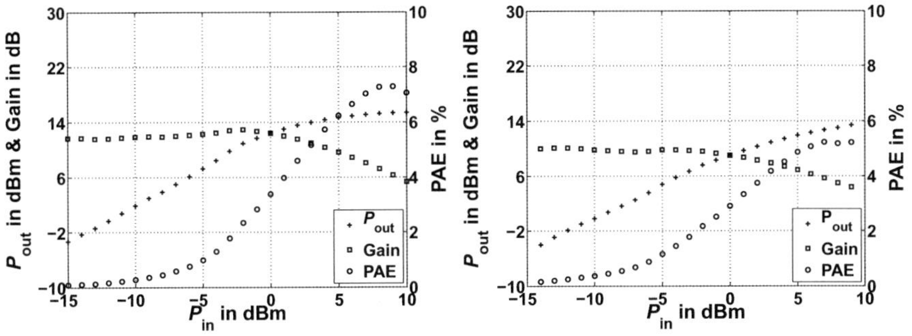

(a) 90 GHz (b) 140 GHz

Figure 4.16: Measured large-signal parameters of the traveling wave amplifier at 90 GHz and 140 GHz. ($I_D = 500\,mA/mm$, $V_D = 1.8\,V$)

110 GHz the available input power was limited, which is why there is only one more frequency point at 140 GHz, where an input power of 8 dBm could be provided in the measurement. To demonstrate the general behavior of the output power versus frequency above 110 GHz, Fig. 4.17(b) demonstrates the measured output power for an input power of 8 dBm and 3 dBm in 10 GHz steps. If the measurement points for certain frequencies are missing, the input power mentioned in the legend of the plot, could not be provided. As visible, the PA has an output power of about 15 dBm up to 110 GHz. Beyond 110 GHz the output power drops slightly. At 140 GHz the PA still has a measured output power of 13 dBm.
The maximum measured linear output power at the -1 dB compression point is plotted in Fig. 4.18. Up to 120 GHz the PA's linear output power is more than 10 dBm. At 140 GHz the 1 dB compression point output power is still 9.7 dBm.

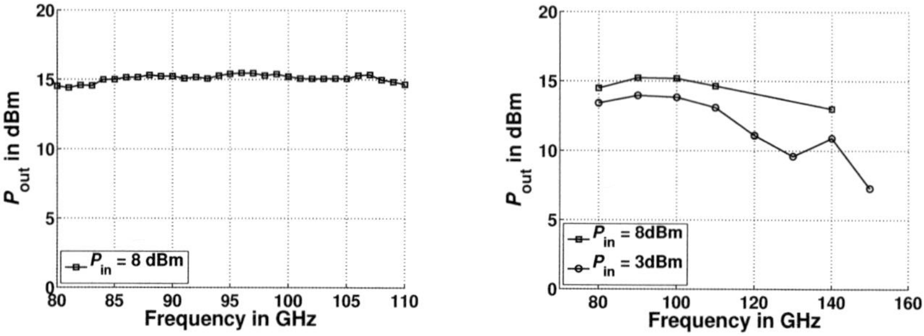

(a) Output power versus frequency from 80 GHz to 110 GHz

(b) Measured output power versus frequency from 80 GHz to 150 GHz

Figure 4.17: Output power vs. frequency of the TWA. ($I_D = 500\,\mathrm{mA/mm}, V_D = 1.8\,\mathrm{V}$)

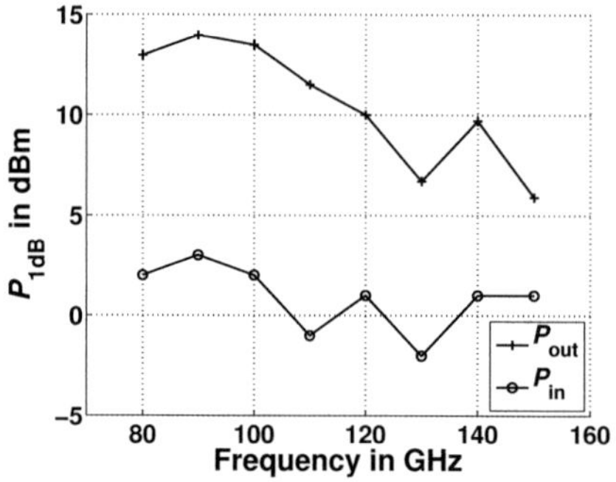

Figure 4.18: Measured 1 dB compression point versus frequency of the traveling wave amplifier. ($I_D = 500\,\mathrm{mA/mm}$, $V_D = 1.8\,\mathrm{V}$)

4.3 Conclusion on chapter 4

In the last chapter of this thesis the applicability of the DT concept for the design of very broadband mmW PAs has been demonstrated. The first PA was designed for the frequency range between 200 GHz and 270 GHz and uses a 1:2 DT (3 port) at its input and output in conjunction with broadband low impedance gain cells based on LC networks. The PA demonstrates a measured small-signal gain of 29 dB at 237 GHz and a 3 dB bandwidth of 69 GHz (201 GHz - 270 GHz), i.e. a relative bandwidth of 29 %. Simulations and measurements of the gain cell and the DT of the PA demonstrate that the PA's bandwidth is mainly limited by the gain cell ($BW_{rel.} = 30\,\%$). If gain cells with a wider bandwidth would have been used, the bandwidth of the DT ($BW_{rel.} \geq 100\,\%$) could have easily supported them, which in return would have led to a wider PA bandwidth.
In terms of output power the PA demonstrated 4.9 dBm at 260 GHz, which corresponds to a power density of 77 $\frac{\text{mW}}{\text{mm}}$. Compared to other published PAs on the same technology, the power density is comparable but the bandwidth of the PA presented in this thesis is much wider.
Besides the impedance transformation, the gain cell design also makes use of the differential signals provided by the DTs. The DC biasing in the presented PA is done through parallel stubs of the LC networks of the gain cells. By directly connecting the ends of adjacent stubs of the two parallel branches the DC power is provided to all gain cells. Due to the differential signals, the connection of the stubs results in vGNDs at their intersection which behave like RF shorts, i.e huge parallel capacitors. Unlike the vGND, however, a huge capacitor would occupy a notable amount of wafer space. The DT therefore facilitates a more compact design.
The second PA presented in this chapter is based on two parallel TWA gain cells in combination with another 1:2 DT (3 port). In this PA, the DT does not only provide two differential 25 Ω ports but also the DC bias for the gain cells. By connecting the DC supply line to the vGND point in the middle of the slab inductor of the differential side of the DT, the DC connection does not affect the RF behavior of the PA but provides the necessary DC power in an efficient and elegant way. Since the TWA gain cells are not biased through the resistors used to terminate the artificial transmission lines, the PA achieves a relatively high PAE for a TWA (7.3 % at 90 GHz).

The broadband impedance transformation of the DT enables the TWA gain cells to use artificial transmission lines with a characteristic impedance of 25 Ω. As explained in the chapter, this leads to an increase of the upper frequency limit of the gain cell and enables the operation of the PA up to 146 GHz. The comparison of the bandwidth of the DT ($BW_{rel.} = 94\,\%$) and the gain cell ($BW_{rel.} = 176\,\%$) makes clear that unlike for the first presented PA, in this case the DT is the limiting element with respect to the BW of the PA ($BW_{rel.} = 98\,\%$). Over the entire 3 dB bandwidth (i.e. 50 GHz - 146 GHz), the PA has an average small-signal gain of 10.8 dB and a peak gain of 12 dB at 106.5 GHz.
Up to 110 GHz the measured output power of the PA is about 15 dBm and up to 140 GHz the maximum output power is still more than 13 dBm.
Both PAs described in this chapter extend the state of the art in terms of bandwidth for mmW PAs. In Fig. 4.19 the state of the art plot shown in chapter 1 is completed and extended by both PAs presented in this chapter. While both PAs achieve competitive output power and PAE levels their BW is decisively higher than what was the current state of the art before this thesis.

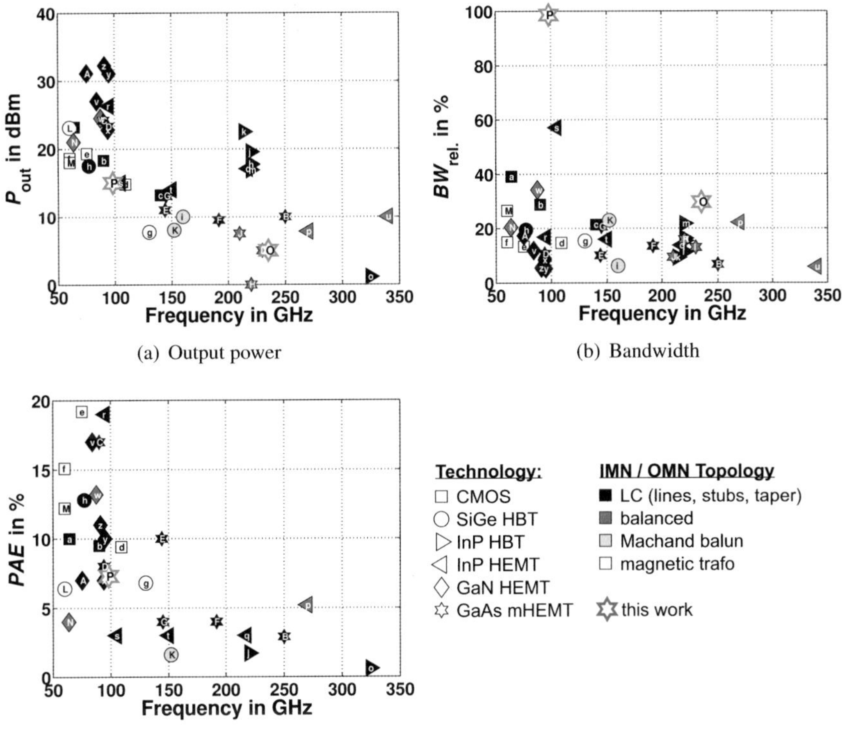

(a) Output power

(b) Bandwidth

(c) PAE

Figure 4.19: Comparison between the state of the art in millimeter-wave power amplifier design and the two amplifiers presented in this work. (O: low impedance PA, P: TWA PA)

5 Conclusion and Discussion

Today, MMIC based mmW communication and radar systems already demonstrate very high data rates and resolutions. One of the bottlenecks of all published MMIC communication and radar front-ends however is the available output power and bandwidth provided by the last stage of the transmitter: the power amplifier.

The main design objective for a broadband PA is the design of a broadband and efficient power divider/combiner matching network. On III-V MMIC technologies almost exclusively LC networks and couplers are used for this purpose. However, the efficiency and/or bandwidth of LC networks degrades, if the required impedance transformation ratio increases. Usually this ratio is already quite high, if just a single transistor of an advanced MMIC process, operated at high mmW frequencies, has to be matched to the common $50\,\Omega$ system impedance. In a PA, on the other hand, usually multiple transistors are used in parallel to overcome the power limitation of a single transistor device. For an LC network this means that the impedance ratio becomes even higher and therefore the efficiency and/or bandwidth is even lower.

Due to the higher losses of passive elements (e.g. transmission lines, capacitors, inductors, etc.) on Si based MMIC technologies, the issue of the low efficiency of LC networks is even more severe than on III-V processes. Literature reveals that highly efficient PAs on Si use a different matching method based on magnetic transformers. The advantage of a magnetic transformer based PA matching network compared to an LC network is that in a first approximation, its efficiency and BW are independent of the required impedance transformation ratio. Furthermore, the operation of multiple low impedance transistor devices plays in favor of the prevailing type of magnetic transformer used for Si based PA MMICs, the DAT/DT, whose impedance transformation ratio increases with the number of parallel ports.

Since III-V based PAs operated at high mmW frequencies are also limited in their BW and/or efficiency, due to the limitations of LC networks, this thesis has investigated the possibility of designing efficient and broadband PAs on III-V technologies based on the concept of DT/DAT.

Besides the motivation and definition of the objective of this thesis, chapter 1 briefly explained the principle concept behind a DT/DAT. Based on this short introduction chapter 2 started the theoretical analysis of the DT by dividing the transformer's ECM into a number of cascaded DT unit cells. Originating from this unit cell, a simple lumped ECM usually used in literature to model a DT/DAT has been developed. Optimization equations from literature, which are based on the simple lumped ECM, have been revised. Their application on the lumped ECM has revealed that the optimization equations for maximum efficiency do not result in the maximum symmetrical bandwidth simultaneously. Therefore, novel design equations for the design of DTs with a maximum symmetrical bandwidth, for a given set of transformer properties and IL or RL condition, have been developed. In general the DT behaves like a second order bandpass filter, which can be optimized to show the standard filter responses. Maximally flat (i.e. Butterworth) frequency response and equal ripple (i.e. Chebyshev) frequency responses, for a given minimum RL, have been presented.

To verify the applicability of the design equations for maximum efficiency and maximum symmetrical bandwidth at high mmW frequencies, a really distributed ECM has been presented that applies coupled lines instead of a lumped transformer to model the coupling between the primary and the secondary side of the DT. Due to its distributed behavior, the new model is more realistic than the lumped model, but in return also more complex. With the new model, the newly developed optimization equations for maximum symmetrical BW could be verified. However, the optimization equations for maximum efficiency, used for the design of DT/DAT in literature, failed to optimize the distributed ECM and therefore would fail to optimize a real DT/DAT at mmW frequencies.

From an evaluation of per unit length quantities (R', L', G', C') of a lossy coupled line, it was found that such lines have a non-negligible resistive coupling between each other. Thus, a new lumped ECM of the DT unit cell has been developed that incorporates the resistive coupling. Its ability to model the efficiency of the distributed ECM of a DT correctly has been verified. Based on the new lumped ECM, new design equations have been developed that are able to optimize the distributed ECM and therefore mmW DTs/DATs correctly.

An assessment of the maximum possible efficiency/minimum IL versus coupling coefficient (k_m) and transformer quality factor (Q) was performed. It revealed the excellent efficiency of the DT, even for relatively small k_m and Q values, which makes the DT/DAT concept realizable on almost every semiconductor process.
After the development of new design equations on the basis of a lumped ECM of a DT unit cell, the chapter ended with the verification that the frequency response of a DT built by a cascade of multiple DT unit cells, equals that of the optimized DT unit cell. It has been found that it is important that the overall DT is small compared to the guided wavelength to guarantee the correct frequency response of the DT and also a good amplitude and phase balance between the ports of a DT consisting of multiple optimized unit cells. Furthermore, it has been shown that the frequency response is almost invariant to the number of unit cells, i.e. the number of DT ports or the impedance transformation ratio. To guarantee that the DT is small compared to the guided wavelength the characteristic impedances of the coupled line, used as transformer in the DT, have to be as high as possible.
As a result of chapter 2 it can be stated that the optimal coupled line for an application in a DT has a low attenuation coefficient (α), resulting in a high quality factor and therefore a high efficiency. It should have a high coupling coefficient to achieve a wide bandwidth and a high efficiency. Furthermore its characteristic impedances should be high, to guarantee that the DT is small compared to the guided wavelength, i.e. the frequency response of the DT equals the optimized frequency response of its DT unit cells.
In chapter 3 the design flow for the development of efficiency or bandwidth optimized DTs has been presented for Fraunhofer IAF's 35 nm and 50 nm mHEMT MMIC technology. A method for the selection of the most suited coupled line type for the two top-metal-layer processes based on the formally determined key parameters (chapter 2) of the coupled line: coupling coefficient, attenuation coefficient and characteristic impedance has been given. The method is based on EM simulations of various coupled line geometries and the extraction of their per unit length quantities. As part of the demonstration of the method, an analytic equation connecting the optimized inductance of the DT unit cell with its length and the four characteristic impedances of an asymmetrical coupled transmission line has been presented for the first time.

From all investigated coupled line types, the asymmetrical coupled line has turned out to be superior for the design of DTs on basis of the IAF MMIC process. Thus, the asymmetrical coupled line type has been used for further investigation.
A DT can be build from multiple unit cells. The way the single unit cells are connected defines the impedance transformation ratio of the DT and the phase relation of the different DT ports. Various combination variations have been presented and the influences of the different wiring options have been explained. For the design of PAs, the voltage combiner type is usually the best DT configuration as it achieves the highest impedance transformation ratio for a given number of ports. Hence, the voltage combiner type DT has been followed up in this thesis.
On Si based technologies usually circular DT geometries are applied to benefit from the DTs ability to provide differential ports. These ports facilitate the use of push-pull transistor pairs using vGNDs between two transistors, which mitigates the loss on lossy Si MMIC technologies decisively. On III-V compound semiconductor processes however, where good/less lossy GND connections are available, a novel DT geometry proposed in this thesis is superior. Compared to the circular geometry, the new straight DT configuration enables a better signal distribution on the MMIC chip by providing a better spacial separation between the ports and avoiding the need for routing signals around the circular DT/DAT geometry. Furthermore, the straight DT geometry influences the EM environment of the single DT unit cells less, which means that the frequency response of the DT and that of its optimized unit cells are more likely to be identical in the straight than the circular DT geometry.
For the two top-metal-layer process used in this thesis, a DT layout has been proposed. It uses the novel straight DT configuration in combination with asymmetrical coupled line type DT unit cells. To design a DT based on this new layout, a design procedure for efficiency or bandwidth optimized DTs has been presented. The approach is solely based on the EM simulation of the asymmetric coupled line geometry, necessary to determine its parameters (i.e. k_m, α, Z_c and Z_π), and the optimization equations of chapter 2. It has been demonstrated that the behavior of an entire optimized 1:4 (5 port) DT can be evaluated only based on S-parameters of the coupled line geometry. To determine the best asymmetric coupled line unit cell in terms of geometry parameters (e.g. line width, ground metal separation, etc.), all possible asymmetrical coupled line geometries for the process have been simu-

lated in an EM field solver. The properties of the optimized DTs, obtained solely from applying the design procedure on each simulated coupled line geometry, have been compared. An extraction of the DT properties from EM simulations of the entire DT layout for each geometry would have taken unrealistically long, due to the high number of different geometry parameter combinations. By knowing the behavior of the DT for all possible coupled line geometries, it is possible to select the coupled line that results in the best trade-off between DT efficiency and bandwidth as well as phase and amplitude balance. To verify the applicability of the design approach, the behavior of multiple optimized DTs, determined solely by S-parameters of the coupled lines and that of full-wave EM simulations of the entire 5 port DTs have been compared. For the three compared configurations a good agreement has been found.

At the end of chapter 3, a comparison between EM simulations and measurements of DT test circuits has been shown. The presented results demonstrate the excellent performance of DTs build by asymmetrical coupled line unit cells in combination with the straight DT geometry implemented by the proposed DT layout on Fraunhofer IAF's mHEMT MMIC technology.

Regarding measurements of DT test structures, the difficulties of the evaluation of the measurement results, due to unwanted impedance transformation caused by the RF-pads, have been discussed and a suitable de-embedding scheme to remedy the issue has been proposed.

The 4th chapter of this thesis finally presented two DT based PAs implemented in Fraunhofer IAF's 35 nm and 50 nm mHEMT MMIC technology. Both PAs have been designed to benefit greatly from the abilities of the DT. They make use of vGNDs, available due to the balun property of the DT, which ease the DC biasing of the gain cells. Furthermore, the gain cells of both PAs benefit from the provided low impedance ports, which enable gain cells with very high bandwidths.

The output power and efficiency of both presented PAs are among the state of the art for mmW PAs at their operation frequency. Their bandwidth, on the other hand, exceeds that of the state of the art by far.

By means of this thesis the state of the art has been extended greatly and the following items shall emphasize this:

- Novel design equations for the design of DT/DAT with maximum symmetrical bandwidth around the design frequency have been developed. Based on DT parameters determined by the DT geometry, the equations enable the design of DTs with a maximally flat or equal ripple insertion loss. As an example, maximally flat IL as well as IL ripples that lead to a RL of 10 dB, 20 dB and 30 dB have been presented.

- It has been demonstrated that the lumped ECM usually used in literature to model the DTs/DATs does not predict the efficiency of DTs/DATs on a lossy MMIC process at mmW frequencies correctly. This results in a wrong efficiency optimization, if the DT/DAT optimization equations from literature, derived from this lumped ECM, are applied. It has been found that the neglect of the severe influence of resistive coupling is the reason why the lumped ECM from literature fails to predict the efficiency correctly.

- A new lumped ECM, which incorporates the resistive coupling and therefore predicts the correct efficiency for a DAT/DT at mmW frequencies has been developed and verified.

- New design equations for the efficiency optimization of DT/DAT at mmW frequencies, based on the new lumped ECM, have been developed and verified.

- For the first time, the parameters of a DT/DAT (efficiency, bandwidth, amplitude and phase balance) and those of the applied coupled lines (i.e. k_m, Z_e, Z_o, Z_c, Z_π and $\frac{l}{\lambda}$) have been put in relation. An analytic equation, which determines the inductance at the two remaining ports of an asymmetrical coupled line, shorted on one side, based on the four characteristic impedances (i.e. $Z_{c,1}$, $Z_{\pi,1}$, $Z_{c,1}$ and $Z_{\pi,1}$) as well as the length of the coupled line (i.e. $\frac{l}{\lambda}$), has been established.

- A novel straight DT geometry has been presented, which shows superior attributes compared to the standard circular DT/DAT geometries widely used in literature. It eases the wiring between the DT/DAT and the active cells of a PA and shows less unwanted influence on the optimized DT/DAT unit cells.

- This thesis proposed a new design strategy for DTs/DATs on MMIC processes. Based on all findings above, it has been shown that the evaluation of the best DT/DAT coupled line geometry for a certain trade-off between efficiency, bandwidth, amplitude and phase balance can be performed just from evaluating the EM simulations of all possible coupled line geometries.
- Two DT based PAs at high mmW frequencies have been presented. The two PAs apply the new concepts, which have been proposed and developed in this thesis. Both PAs extend the current state of the art in terms of relative bandwidth at their operation frequencies, while showing competitive output power and efficiency. Furthermore, the PA operating at 200 GHz - 270 GHz is the highest DT/DAT based PA ever reported.

The findings presented in this thesis facilitate the design of ultra broadband impedance transforming power combiner/divider circuits based on the principle of magnetic transformers. While the focus of this work was on the design of mmW PAs, there are other circuits like frequency multipliers and phase shifters, which can also benefit from using DTs and therefore from the results of this thesis. The key factors for radar and communication systems at mmW frequencies are bandwidth and signal power, which translate into high data rates and high radar resolutions. Both, the signal power and the bandwidth of MMIC based systems, are mainly determined by the efficiency and bandwidth of the matching networks applied in every part of those MMICs. Hence, by providing insight into suitable matching techniques and by developing methods for their optimization, one provides the means to improve not only single RF circuits but also to leverage all sorts of mmW systems. This thesis does this for the concept of the distributed transformer, which is believed to play an important role as matching network in all sorts of MMIC based systems in the future.

A State of the Art in MMW PAs - Reference List

ID	Reference	ID	Reference	ID	Reference
a	[HTL+13]	b	[HTL+13]	c	[HTL+13]
d	[GXC12]	e	[WCW12]	f	[CN11]
g	[HXG+12]	h	[KH06]	i	[SCP13]
j	[RRG+12a]	k	[RGR+13]	l	[GRRF13]
m	[RRG+11]	n	[RRG+12b]	o	[HUM+08]
p	[DMR+08]	q	[RLM+10]	r	[CIL+98]
s	[SL01]	t	[SPH+04]	u	[RDL+10]
v	[MKM+08]	w	[MKM+12]	x	[QTK+11]
y	[BBC+11]	z	[BBC+11]	A	[NMM+10]
B	[Die14]	C	[HBR+05]	D	[TLSM05]
E	[KPM+09a]	F	[KPM+09b]	G	[DPG+12]
H	[LDM+12]	I	[LDM+13]	J	[LDM+13]
K	[SHP13]	L	[PG07]	M	[JTHW09]
N	[PDS+13]	O	this work	P	this work

Table A.1: Reference list for the state of the art plots in Fig. 1.3, 4.19 and A.1.

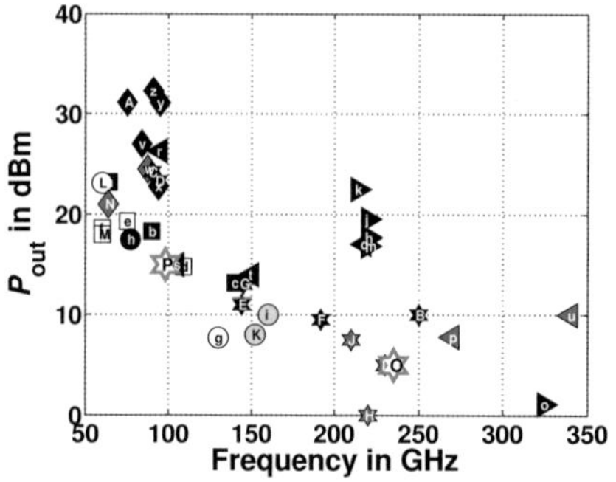

(a) Output power

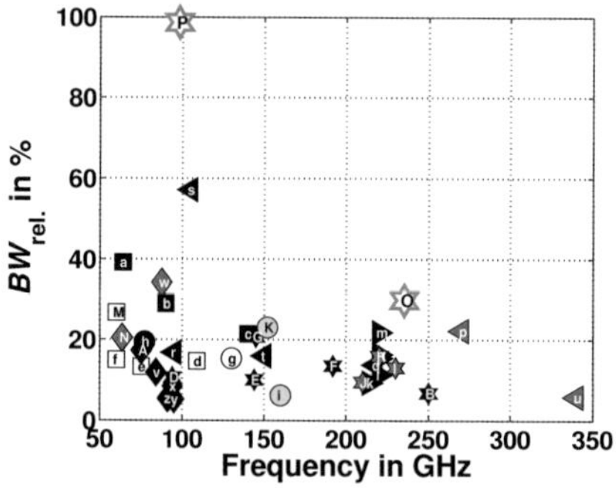

(b) Bandwidth

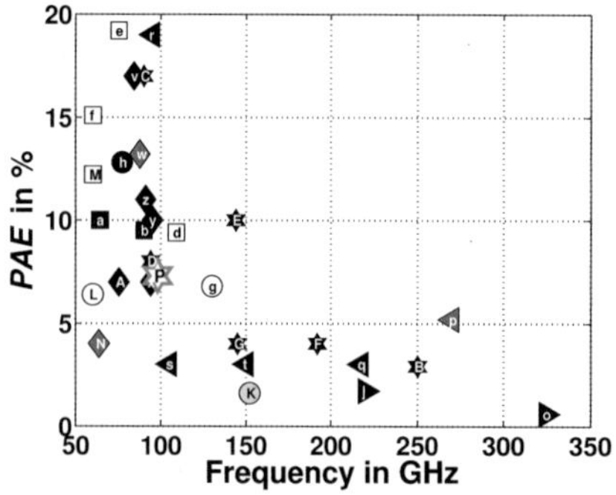

(c) PAE

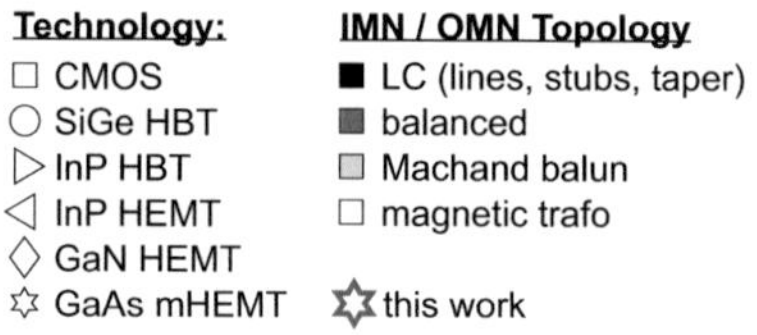

Figure A.1: State of the art in millimeter-wave power amplifier design, including the results of this work.

B State of the Art in MMW DAT and DT PAs - Reference List

ID	Reference	ID	Reference	ID	Reference
a	[AKRH01]	b	[PGFR05]	c	[PG07]
d	[PHKH08]	e	[CRN08]	f	[HCR+08]
g	[LPH09]	h	[CRN09]	i	[MKS+10]
j	[LVG10]	k	[HLR10]	l	[EMPvR11]
m	[CN11]	n	[KBH11]	o	[ALD+12]
p	[WCW12]	q	[GXC12]	r	[OKH13]
s	[HXG+12]	t	[JTHW09]	u	[TTB+13]
v	[PKKH07]	w	[HTLW10]	x	[LTLW07]
y	this work	z	this work		

Table B.1: Reference list for the state of the art plots in Fig. 1.9 and B.1.

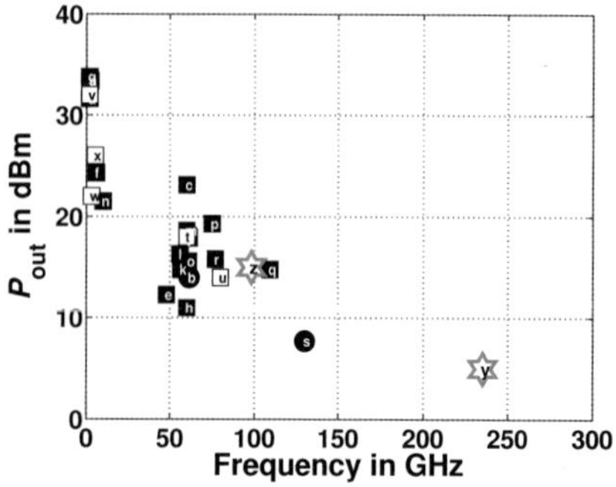

(a) Output power

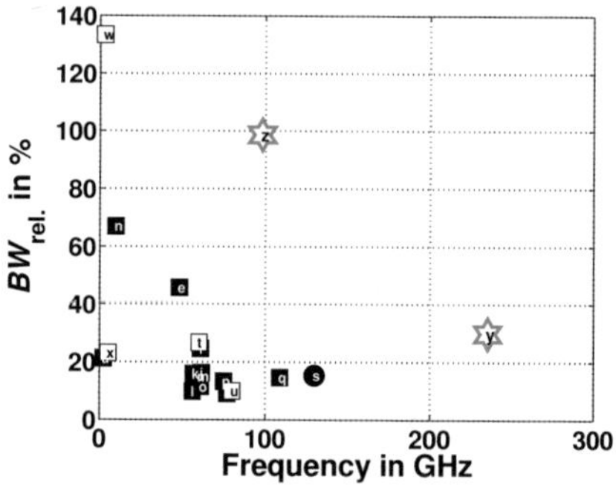

(b) Bandwidth

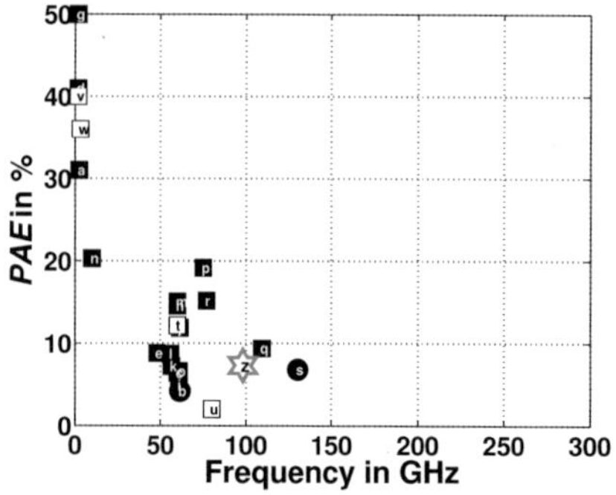

(c) PAE

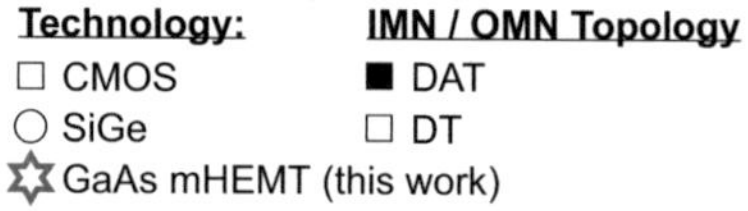

Figure B.1: State of the art in magnetic transformer based millimeter-wave power amplifier design.

C 4-Port S- to ABCD-Matrix Conversion

The derivation of a 2N port ABCD matrix in terms of the S-parameter matrix is mainly based on [LCJ$^+$03]. To define an equation transforming the S-Parameter matrix into an ABCD-matrix, a relation between the voltages and currents as well as the power waves at the different ports has to be defined. Figure C.1(a) shows the definition of voltages (V) and currents (I) at the different ports of a 2N-port, while figure C.1(b) does the same for the injected (a) and reflected (b) power waves.

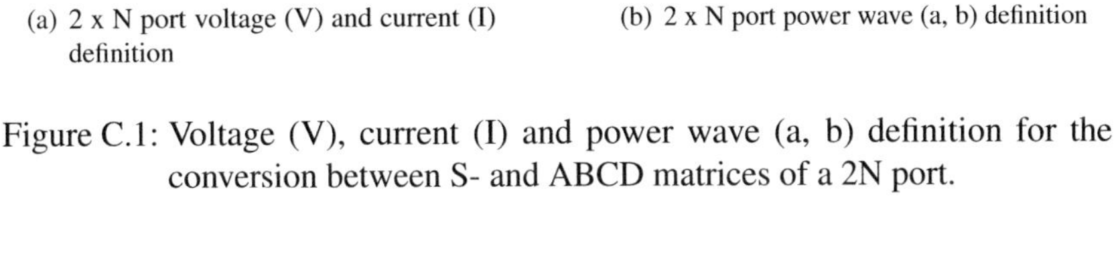

(a) 2 x N port voltage (V) and current (I) definition

(b) 2 x N port power wave (a, b) definition

Figure C.1: Voltage (V), current (I) and power wave (a, b) definition for the conversion between S- and ABCD matrices of a 2N port.

In this thesis most of the networks have four or less ports. Hence, for the sake of clarity, the following equations will be derived for a four port instead of a cumbersome general 2N port solution.

The ABCD-matrix, which shall be derived in terms of S-Parameters and port impedances can be expressed like this:

$$\begin{bmatrix} \overline{\overline{U_1}} \\ \overline{\overline{I_1}} \end{bmatrix} = \begin{bmatrix} A & B \\ C & D \end{bmatrix} \cdot \begin{bmatrix} \overline{\overline{U_2}} \\ -\overline{\overline{I_2}} \end{bmatrix} \tag{C.1}$$

where $\overline{\overline{U_1}} = \begin{bmatrix} U_1 \\ U_2 \end{bmatrix}$ $\overline{\overline{U_2}} = \begin{bmatrix} U_3 \\ U_4 \end{bmatrix}$ $\overline{\overline{I_1}} = \begin{bmatrix} I_1 \\ I_2 \end{bmatrix}$ $\overline{\overline{I_2}} = \begin{bmatrix} I_3 \\ I_4 \end{bmatrix}$ and A, B, C and D are 2×2 matrices.

To combine the ABCD matrix with its S-Parameter counterpart, the power waves can be defined in terms of the voltages and currents at the various ports in the same way as for the ABCD matrix. Equation C.2 and C.3 show the definition in matrix notation used throughout this thesis for the b and a waves respectively:

$$\begin{aligned} \begin{bmatrix} \overline{\overline{b_1}} \\ \overline{\overline{b_2}} \end{bmatrix} &= \frac{1}{2}\left(\overline{\overline{\Gamma}} \begin{bmatrix} \overline{\overline{U_1}} \\ \overline{\overline{I_1}} \end{bmatrix} + \overline{\overline{\Theta}} \begin{bmatrix} \overline{\overline{U_2}} \\ -\overline{\overline{I_2}} \end{bmatrix} \right) \\ &= \frac{1}{2}\left(\begin{bmatrix} \overline{\overline{I}} & -\overline{\overline{Z_1}} \\ 0 & 0 \end{bmatrix} \begin{bmatrix} \overline{\overline{U_1}} \\ \overline{\overline{I_1}} \end{bmatrix} + \begin{bmatrix} 0 & 0 \\ \overline{\overline{I}} & \overline{\overline{Z_2}} \end{bmatrix} \begin{bmatrix} \overline{\overline{U_2}} \\ -\overline{\overline{I_2}} \end{bmatrix} \right) \end{aligned} \tag{C.2}$$

$$\begin{aligned} \begin{bmatrix} \overline{\overline{a_1}} \\ \overline{\overline{a_2}} \end{bmatrix} &= \frac{1}{2}\left(\overline{\overline{\Phi}} \begin{bmatrix} \overline{\overline{U_1}} \\ \overline{\overline{I_1}} \end{bmatrix} + \overline{\overline{\Psi}} \begin{bmatrix} \overline{\overline{U_2}} \\ -\overline{\overline{I_2}} \end{bmatrix} \right) \\ &= \frac{1}{2}\left(\begin{bmatrix} \overline{\overline{I}} & \overline{\overline{Z_1}} \\ 0 & 0 \end{bmatrix} \begin{bmatrix} \overline{\overline{U_1}} \\ \overline{\overline{I_1}} \end{bmatrix} + \begin{bmatrix} 0 & 0 \\ \overline{\overline{I}} & -\overline{\overline{Z_2}} \end{bmatrix} \begin{bmatrix} \overline{\overline{U_2}} \\ -\overline{\overline{I_2}} \end{bmatrix} \right) \end{aligned} \tag{C.3}$$

where

$$\overline{\overline{a_1}} = \begin{bmatrix} a_1 \\ a_2 \end{bmatrix} \quad \overline{\overline{a_2}} = \begin{bmatrix} a_3 \\ a_4 \end{bmatrix} \quad \overline{\overline{b_1}} = \begin{bmatrix} b_1 \\ b_2 \end{bmatrix} \quad \overline{\overline{b_2}} = \begin{bmatrix} b_3 \\ b_4 \end{bmatrix}$$

$$\overline{\overline{Z_1}} = \begin{bmatrix} Z_1 & 0 \\ 0 & Z_2 \end{bmatrix} \quad \overline{\overline{Z_2}} = \begin{bmatrix} Z_3 & 0 \\ 0 & Z_4 \end{bmatrix} \quad \overline{\overline{I}} = \begin{bmatrix} 1 & 0 \\ 0 & 1 \end{bmatrix}$$

Substituting these equations in the standard S-parameter definition

$$\begin{bmatrix} b_1 \\ b_2 \\ b_3 \\ b_4 \end{bmatrix} = \begin{bmatrix} S_{11} & S_{12} & S_{13} & S_{14} \\ S_{21} & S_{22} & S_{23} & S_{24} \\ S_{31} & S_{32} & S_{33} & S_{34} \\ S_{41} & S_{42} & S_{43} & S_{44} \end{bmatrix} \begin{bmatrix} a_1 \\ a_2 \\ a_3 \\ a_4 \end{bmatrix}$$
$$\hat{=} \begin{bmatrix} \overline{\overline{b_1}} \\ \overline{\overline{b_2}} \end{bmatrix} = \begin{bmatrix} \overline{\overline{S_{11}}} & \overline{\overline{S_{12}}} \\ \overline{\overline{S_{21}}} & \overline{\overline{S_{22}}} \end{bmatrix} \begin{bmatrix} \overline{\overline{a_1}} \\ \overline{\overline{a_2}} \end{bmatrix}$$
$$\hat{=} \begin{bmatrix} \overline{\overline{b_1}} \\ \overline{\overline{b_2}} \end{bmatrix} = \overline{\overline{S}} \begin{bmatrix} \overline{\overline{a_1}} \\ \overline{\overline{a_2}} \end{bmatrix}$$

results in the following equation:

$$\left(\overline{\overline{\Gamma}} \begin{bmatrix} \overline{\overline{U_1}} \\ \overline{\overline{I_1}} \end{bmatrix} + \overline{\overline{\Theta}} \begin{bmatrix} \overline{\overline{U_2}} \\ -\overline{\overline{I_2}} \end{bmatrix} \right) = \overline{\overline{S}} \left(\overline{\overline{\Phi}} \begin{bmatrix} \overline{\overline{U_1}} \\ \overline{\overline{I_1}} \end{bmatrix} + \overline{\overline{\Psi}} \begin{bmatrix} \overline{\overline{U_2}} \\ -\overline{\overline{I_2}} \end{bmatrix} \right) \tag{C.4}$$

By solving his matrix equation for $\overline{\overline{\mathrm{V}_1}}$ the ABCD-matrix representation can be obtained:

$$\begin{bmatrix} \overline{\overline{U_1}} \\ \overline{\overline{I_1}} \end{bmatrix} = \left(\overline{\overline{\Gamma}} - \overline{\overline{S\Phi}} \right)^{-1} \left(\overline{\overline{S\Psi}} - \overline{\overline{\Theta}} \right) \begin{bmatrix} \overline{\overline{U_2}} \\ -\overline{\overline{I_2}} \end{bmatrix} \tag{C.5}$$

Rewriting this equation C.5 in terms of S-parameters, port impedances as well as voltages and currents at the ports leads to a matrix equation, which derives the 2N-port ABCD-parameters from 2N-port S-parameters:

$$\begin{bmatrix} A & B \\ C & D \end{bmatrix} = \begin{bmatrix} I - \overline{\overline{S_{11}}} & -Z_0(\overline{\overline{I}} + \overline{\overline{S_{11}}}) \\ \overline{\overline{S_{21}}} & Z_0\overline{\overline{S_{11}}} \end{bmatrix}^{-1} \times \begin{bmatrix} \overline{\overline{S_{12}}} & -Z_0\overline{\overline{S_{12}}} \\ \overline{\overline{I}} - \overline{\overline{S_{22}}} & Z_0(I + \overline{\overline{S_{22}}}) \end{bmatrix} \tag{C.6}$$

D Derivation of the Efficiency of a Coupled Line 1:1 Transformer

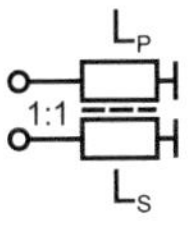

(a) Single coupled line element.

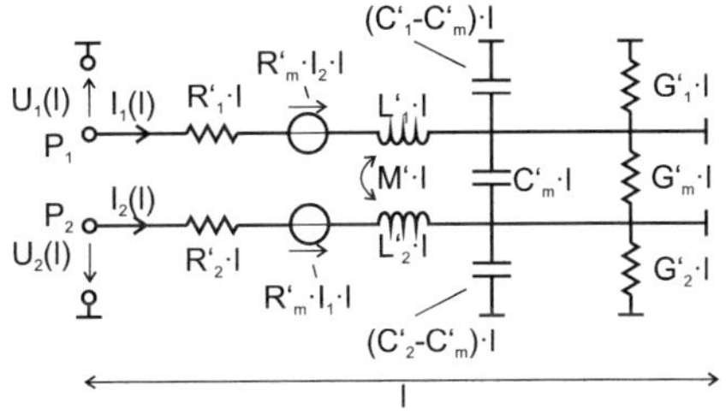

(b) Coupled line ECM connected as DT (shorted)

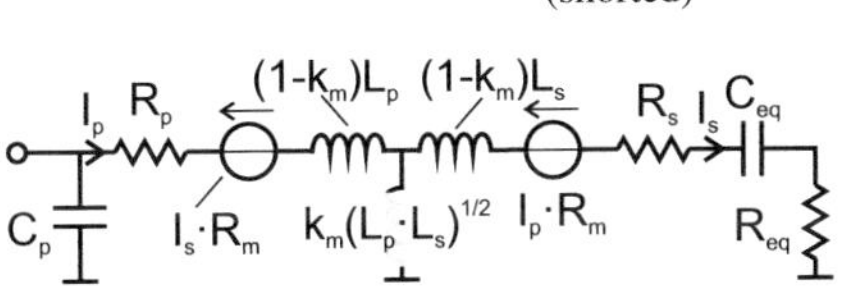

(c) Schematic of a DT

Figure D.1: The equivalent lumped coupled line model used for the efficiency calculations of DT.

Within this section the efficiency of the DT will be derived:

$$
\begin{aligned}
\eta &= \frac{P_{\text{out}}}{P_{\text{out}} + P_{\text{diss}}} \\
&= \frac{|I_{\text{s}}|^2 \cdot R_{\text{eq}}}{|I_{\text{s}}|^2 \cdot R_{\text{eq}} + |I_{\text{s}}|^2 \cdot R_{\text{s}} + |I_{\text{p}}|^2 \cdot R_{\text{p}} - (I_{\text{s}} \cdot R_{\text{m}}) \cdot I_{\text{p}}^* - (I_{\text{p}} \cdot R_{\text{m}}) \cdot I_{\text{s}}^*} \\
&= \frac{R_{\text{eq}}}{R_{\text{eq}} + R_{\text{s}} + \frac{|I_{\text{p}}|^2}{|I_{\text{s}}|^2} \cdot R_{\text{p}} - 2\text{Re}\left(\frac{I_{\text{p}}}{I_{\text{s}}}\right) \cdot R_{\text{m}}}
\end{aligned} \tag{D.1}
$$

where the relation between the current on the primary (I_{p}) and secondary (I_{s}) side of the DT is as follows:

$$
\begin{aligned}
\frac{I_{\text{p}}}{I_{\text{s}}} &= \frac{j\omega(1-k_{\text{m}})L_s + R_s + R_{eq} + \frac{1}{j\omega C_{eq}} - \frac{I_p R_m}{I_s}}{j\omega k_{\text{m}}\sqrt{L_{\text{p}}L_{\text{s}}} || \left[j\omega(1-k_{\text{m}})L_{\text{s}} + R_{\text{s}} + R_{\text{eq}} + \frac{1}{j\omega C_{\text{eq}}} - \frac{I_{\text{p}}R_{\text{m}}}{I_{\text{s}}} \right]} \\
&= \frac{j\omega(1-k_{\text{m}})L_{\text{s}} + R_{\text{s}} + R_{\text{eq}} + \frac{1}{j\omega C_{\text{eq}}} - \frac{I_{\text{p}}R_{\text{m}}}{I_{\text{s}}} + j\omega k_{\text{m}}\sqrt{L_{\text{p}}L_{\text{s}}}}{j\omega k_{\text{m}}\sqrt{L_{\text{p}}L_{\text{s}}}}
\end{aligned} \tag{D.2}
$$

Since the primary and secondary inductance of the DT are almost identical ($L = L_{\text{p}} \approx L_{\text{s}}$), Eqn. D.2 can be further simplified:

$$
\begin{aligned}
\frac{I_{\text{p}}}{I_{\text{s}}} &= \frac{j\omega L + R_{\text{s}} + R_{\text{eq}} + \frac{1}{j\omega C_{\text{eq}}} - \frac{I_{\text{p}}R_{\text{m}}}{I_{\text{s}}}}{j\omega k_{\text{m}} L} \\
\frac{I_{\text{p}}}{I_{\text{s}}}\left(1 + \frac{R_{\text{m}}}{j\omega k_{\text{m}} L}\right) &= \frac{j\omega L + R_{\text{s}} + R_{\text{eq}} + \frac{1}{j\omega C_{\text{eq}}}}{j\omega k_{\text{m}} L} \\
\frac{I_{\text{p}}}{I_{\text{s}}} &= \frac{j\omega L + R_{\text{s}} + R_{\text{eq}} + \frac{1}{j\omega C_{\text{eq}}}}{j\omega k_{\text{m}} L + R_{\text{m}}}
\end{aligned} \tag{D.3}
$$

Therefore by substituting Eqn. D.3 in Eqn. D.1 the efficiency of a single DT coupled line element can be defined as follows:

$$\eta = \frac{R_{\mathrm{eq}}}{R_{\mathrm{eq}} + R_{\mathrm{s}} + \left| \frac{j\omega L + R_{\mathrm{s}} + R_{\mathrm{eq}} + \frac{1}{j\omega C_{\mathrm{eq}}}}{j\omega k_{\mathrm{m}} L + R_{\mathrm{m}}} \right|^2 \cdot R_{\mathrm{p}} - 2\,\mathrm{Re}\left\{ \frac{j\omega L + R_{\mathrm{s}} + R_{\mathrm{eq}} + \frac{1}{j\omega C_{\mathrm{eq}}}}{j\omega k_{\mathrm{m}} L + R_{\mathrm{m}}} \right\} \cdot R_{\mathrm{m}}} \tag{D.4}$$

By replacing R_{p}, R_{s} and R_{m} in Eqn. D.4 by the quality factors of the primary, secondary and mutual inductance as well as the inductance L (Eqn. D.5)

$$\begin{aligned} Q_{\mathrm{p}} &= \frac{\omega L_{\mathrm{p}}}{R_{\mathrm{p}}} \approx \frac{\omega L}{R_{\mathrm{p}}} \\ Q_{\mathrm{s}} &= \frac{\omega L_{\mathrm{s}}}{R_{\mathrm{s}}} \overset{L_{\mathrm{s}} \approx L_{\mathrm{p}} = L}{=} \frac{\omega L}{R_{\mathrm{s}}} \\ Q_{\mathrm{m}} &= \frac{\omega M}{R_{\mathrm{m}}} = \frac{\omega k_{\mathrm{m}} \sqrt{L_{\mathrm{p}} L_{\mathrm{s}}}}{R_{\mathrm{m}}} \overset{L_{\mathrm{s}} \approx L_{\mathrm{p}} = L}{=} \frac{\omega k_{\mathrm{m}} L}{R_{\mathrm{m}}} \end{aligned} \tag{D.5}$$

Eqn. 2.26 from chapter 2 is obtained:

$$\eta = \frac{R_{\mathrm{eq}}}{R_{\mathrm{eq}} + \frac{\omega L}{Q_{\mathrm{s}}} + \left| \frac{j\omega L + \frac{\omega L}{Q_{\mathrm{s}}} + R_{\mathrm{eq}} + \frac{1}{j\omega C_{\mathrm{eq}}}}{j\omega k_{\mathrm{m}} L + \frac{\omega k_{\mathrm{m}} L}{Q_{\mathrm{m}}}} \right|^2 \cdot \frac{\omega L}{Q_{\mathrm{p}}} - 2\,\mathrm{Re}\left\{ \frac{j\omega L + \frac{\omega L}{Q_{\mathrm{s}}} + R_{\mathrm{eq}} + \frac{1}{j\omega C_{\mathrm{eq}}}}{j\omega k_{\mathrm{m}} L + \frac{\omega L}{Q_{\mathrm{m}}}} \right\} \cdot \frac{\omega k_{\mathrm{m}} L}{Q_{\mathrm{m}}}} \tag{D.6}$$

E Coupled Line Theory

This appendix is mostly based on [Tri75, MBBH07].

The voltages and currents of two coupled lines can be expressed by the following four equations:

$$-\frac{dU_1}{dx} = z_1 I_1 + z_m I_2 \tag{E.1}$$

$$-\frac{dU_2}{dx} = z_2 I_2 + z_m I_1 \tag{E.2}$$

$$-\frac{dI_1}{dx} = y_1 U_1 + y_m U_2 \tag{E.3}$$

$$-\frac{dI_2}{dx} = y_2 U_2 + y_m U_1 \tag{E.4}$$

z_1, z_2, z_m are the self and mutual impedances per unit length of either line in the presence of the other line.

y_1, y_2, y_m are the self and mutual admittances per unit length of either line in the presence of the other line.

In this appendix a harmonic time variation with $e^{j\omega t}$ is assumed.

By differentiating equation E.1 and E.2 with respect to the position x and by substituting E.3 and E.4, a system of equations for the voltages on the uniformly coupled lines is obtained:

$$\frac{d^2U_1}{dx^2} - a_1 U_1 - b_1 U_2 = 0 \tag{E.5}$$

$$\frac{d^2U_2}{dx^2} - a_2 U_2 - b_2 U_1 = 0 \tag{E.6}$$

where

$$\begin{aligned} a_1 &= y_1 z_1 + y_m z_m \\ a_2 &= y_2 z_2 + y_m z_m \\ b_1 &= z_1 y_m + y_2 z_m \\ b_2 &= z_2 y_m + y_1 z_m \end{aligned} \tag{E.7}$$

Assuming an x variation of $U(x) = U_0 e^{\gamma x}$ for the voltages on the lines, the solution of the eigenvalue problem leads to the following four solutions for the propagation coefficient γ:

$$\begin{aligned} \gamma_{1,2} &= \pm\gamma_c \\ \gamma_{3,4} &= \pm\gamma_\pi \end{aligned}$$

where

$$\gamma_{c,\pi}^2 = \frac{a_1 + a_2}{2} \pm \frac{1}{2}\left[(a_1 - a_2)^2 + 4b_1 b_2\right]^{\frac{1}{2}} \tag{E.8}$$

The relation between the voltages on both lines can be determined by equation E.5, E.6 and E.8 and is given by:

$$\begin{aligned} R_c &\hat{=} \frac{U_2}{U_1} \quad \text{where} \quad \gamma = \pm\gamma_c \\ &= \frac{1}{2b_1}(a_2 - a_1) + \left[(a_1 - a_2)^2 + 4b_1 b_2\right]^{\frac{1}{2}} \end{aligned} \tag{E.9}$$

$$\begin{aligned} R_\pi &\hat{=} \frac{U_2}{U_1} \quad \text{where} \quad \gamma = \pm\gamma_\pi \\ &= \frac{1}{2b_1}(a_2 - a_1) - \left[(a_1 - a_2)^2 + 4b_1 b_2\right]^{\frac{1}{2}} \end{aligned} \tag{E.10}$$

If the coupled lines are identical with respect to their electromagnetic properties R_c and R_π become 1 and -1 respectively, as can be seen by substituting $a_1 = a_2$

and $b_1 = b_2$ in equation E.9 and E.10. By doing this the asymmetric coupled lines' c- and π-modes become the symmetric coupled lines' even- and odd-modes, respectively.

The general solution for the voltages on the two lines in terms of all four propagating waves can be expressed as:

$$\begin{aligned} U_1(x) &= A_1 e^{-\gamma_c x} + A_2 e^{\gamma_c x} + A_3 e^{-\gamma_\pi x} + A_4 e^{\gamma_\pi x} \\ U_2(x) &= A_1 R_c e^{-\gamma_c x} + A_2 R_c e^{\gamma_c x} + A_3 R_\pi e^{-\gamma_\pi x} + A_4 R_\pi e^{\gamma_\pi x} \end{aligned} \tag{E.11}$$

Entering E.11 in E.1 and E.2 the currents on the lines can be expressed as:

$$\begin{aligned} I_1(x) &= \frac{A_1}{Z_{c1}} e^{-\gamma_c x} - \frac{A_2}{Z_{c1}} e^{\gamma_c x} + \frac{A_3}{Z_{\pi 1}} e^{-\gamma_\pi x} - \frac{A_4}{Z_{\pi 1}} e^{\gamma_\pi x} \\ I_2(x) &= \frac{A_1}{Z_{c2}} R_c e^{-\gamma_c x} - \frac{A_2}{Z_{c2}} R_c e^{\gamma_c x} + \frac{A_3}{Z_{\pi 2}} R_\pi e^{-\gamma_\pi x} - \frac{A_4}{Z_{\pi 2}} R_\pi e^{\gamma_\pi x} \end{aligned} \tag{E.12}$$

By doing this, the characteristic impedances for the c and π modes on both lines can be determined in terms of the self and mutual impedances of the two lines as follows:

$$\begin{aligned} Z_{c1} &= \frac{1}{Y_{c1}} = \frac{1}{\gamma_c} \frac{z_1 z_2 - z_m^2}{z_2 - z_m R_c} \\ Z_{c2} &= \frac{1}{Y_{c2}} = \frac{R_c}{\gamma_c} \frac{z_1 z_2 - z_m^2}{z_1 R_c - z_m} \\ Z_{\pi 1} &= \frac{1}{Y_{\pi 1}} = \frac{1}{\gamma_\pi} \frac{z_1 z_2 - z_m^2}{z_2 - z_m R_\pi} \\ Z_{\pi 2} &= \frac{1}{Y_{\pi 2}} = \frac{R_\pi}{\gamma_\pi} \frac{z_1 z_2 - z_m^2}{z_1 R_\pi - z_m} \end{aligned} \tag{E.13}$$

From these equations (E.13) as well as E.9, E.10 it can be seen that for the two modes the following relation between the characteristic impedances/admittances of the lines and the ratio of their voltage waves exists:

$$\frac{Z_{c1}}{Z_{c2}} = \frac{Y_{c2}}{Y_{c1}} = \frac{Z_{\pi 1}}{Z_{\pi 2}} = \frac{Y_{\pi 2}}{Y_{\pi 1}} = -R_c R_\pi \tag{E.14}$$

Equation E.11 and E.12 can be used to obtain the voltages and currents at the four ports of the coupled lines in terms of the voltage amplitudes ($A_1 - A_4$) of the four propagating waves on line 1:

$$\begin{bmatrix} U_1 \\ U_2 \\ U_3 \\ U_4 \end{bmatrix} = \begin{bmatrix} 1 & 1 & 1 & 1 \\ R_c & R_c & R_\pi & R_\pi \\ e^{-\gamma_c l} & e^{\gamma_c l} & e^{-\gamma_\pi l} & e^{\gamma_\pi l} \\ R_c e^{-\gamma_c l} & R_c e^{\gamma_c l} & R_\pi e^{-\gamma_\pi l} & R_\pi e^{\gamma_\pi l} \end{bmatrix} \cdot \begin{bmatrix} A_1 \\ A_2 \\ A_3 \\ A_4 \end{bmatrix} \tag{E.15}$$

$$\begin{bmatrix} I_1 \\ I_2 \\ -I_3 \\ -I_4 \end{bmatrix} = \begin{bmatrix} Y_{c1} & -Y_{c1} & Y_{\pi 1} & -Y_{\pi 1} \\ R_c Y_{c2} & -R_c Y_{c2} & R_\pi Y_{\pi 2} & -R_\pi Y_{\pi 2} \\ Y_{c1} e^{-\gamma_c l} & -Y_{c1} e^{\gamma_c l} & Y_{\pi 1} e^{-\gamma_\pi l} & -Y_{\pi 1} e^{\gamma_\pi l} \\ R_c Y_{c2} e^{-\gamma_c l} & -R_c Y_{c2} e^{\gamma_c l} & R_\pi Y_{\pi 2} e^{-\gamma_\pi l} & -R_\pi Y_{\pi 2} e^{\gamma_\pi l} \end{bmatrix} \cdot \begin{bmatrix} A_1 \\ A_2 \\ A_3 \\ A_4 \end{bmatrix} \tag{E.16}$$

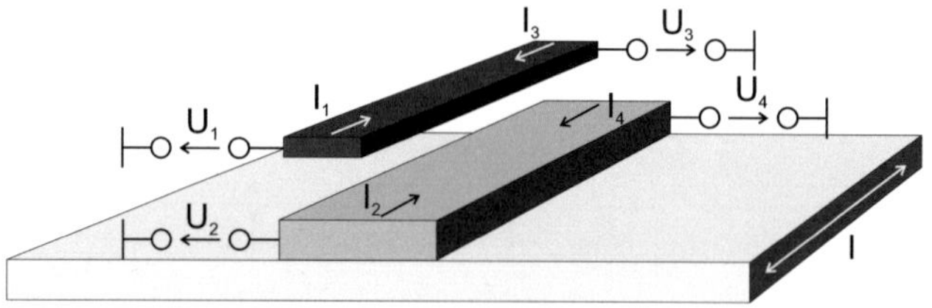

Figure E.1: Schematic of a uniform coupled-line four port

The equations E.15 and E.16 can be combined to obtain the Z- and Y-Parameters of an asymmetrical coupled line. In Eqn. E.17 the Y-Parameters are given:

$$
\begin{aligned}
Y_{11} = Y_{33} &= \frac{Y_{c1}\coth\gamma_c l}{(1 - {}^{R_c}/_{R_\pi})} + \frac{Y_{\pi 1}\coth\gamma_\pi l}{(1 - {}^{R_\pi}/_{R_c})} \qquad \text{(E.17)}\\
Y_{12} = Y_{21} = Y_{34} = Y_{43} &= -\frac{Y_{c1}\coth\gamma_c l}{R_\pi(1 - {}^{R_c}/_{R_\pi})} - \frac{Y_{\pi 1}\coth\gamma_\pi l}{R_c(1 - {}^{R_\pi}/_{R_c})}\\
Y_{14} = Y_{41} = Y_{23} = Y_{32} &= \frac{Y_{c1}}{(R_\pi - R_c)\sinh\gamma_c l} + \frac{Y_{\pi 1}}{(R_c - R_\pi)\sinh\gamma_\pi l}\\
Y_{13} = Y_{31} &= -\frac{Y_{c1}}{(1 - {}^{R_c}/_{R_\pi})\sinh\gamma_c l} - \frac{Y_{\pi 1}}{(1 - {}^{R_\pi}/_{R_c})\sinh\gamma_\pi l}\\
Y_{22} = Y_{44} &= -\frac{R_c Y_{c2}\coth\gamma_c l}{R_\pi(1 - {}^{R_c}/_{R_\pi})} - \frac{R_\pi Y_{\pi 2}\coth\gamma_\pi l}{R_c(1 - {}^{R_\pi}/_{R_c})}\\
Y_{24} = Y_{42} &= \frac{R_c Y_{c2}}{R_\pi(1 - {}^{R_c}/_{R_\pi})\sinh\gamma_c l} + \frac{R_\pi Y_{\pi 2}}{R_c(1 - {}^{R_\pi}/_{R_c})\sinh\gamma_\pi l}
\end{aligned}
$$

F Implementation Results

F.1 Line parameters for three coupled line configurations

F.1.1 Symmetrical coupled line (configuration a)

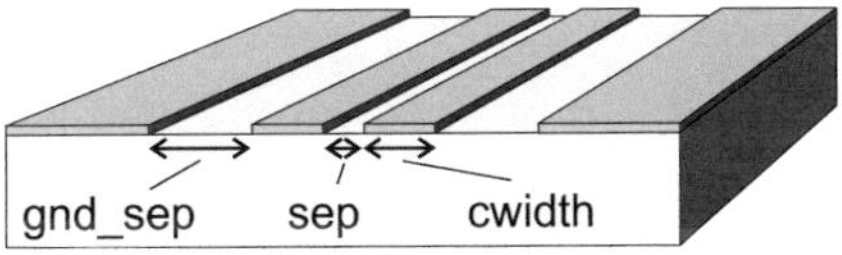

Figure F.1: Geometry of the symmetric coupled line build only with Met1 (configuration a) and realized on IAF's 35 nm and 50 nm process.

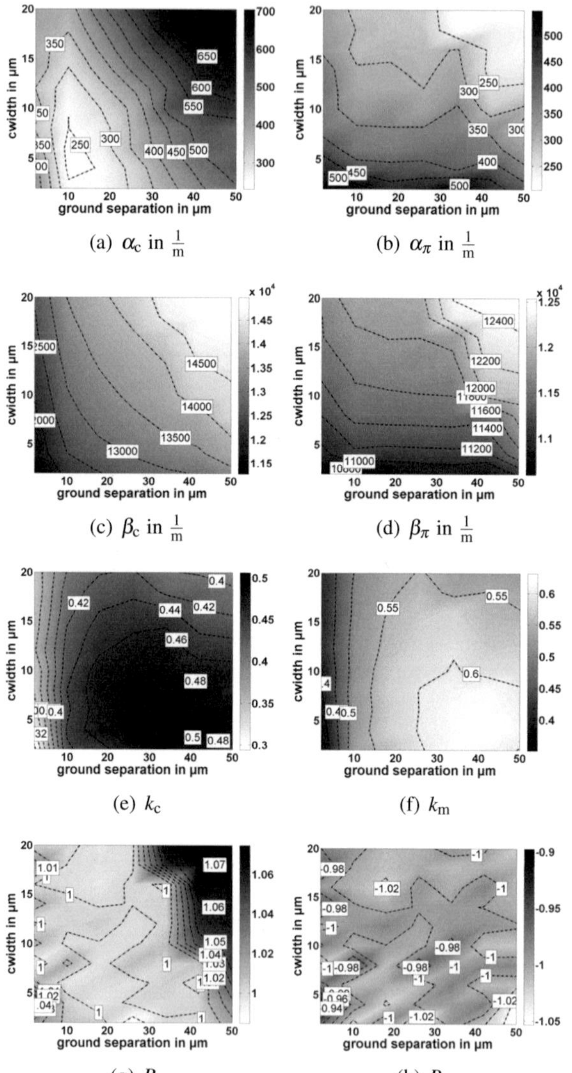

(a) α_c in $\frac{1}{m}$

(b) α_π in $\frac{1}{m}$

(c) β_c in $\frac{1}{m}$

(d) β_π in $\frac{1}{m}$

(e) k_c

(f) k_m

(g) R_c

(h) R_π

Figure F.2: Line parameters of the symmetric coupled line build only with Met1 (configuration a) at 240 GHz. (sep = 2 µm)

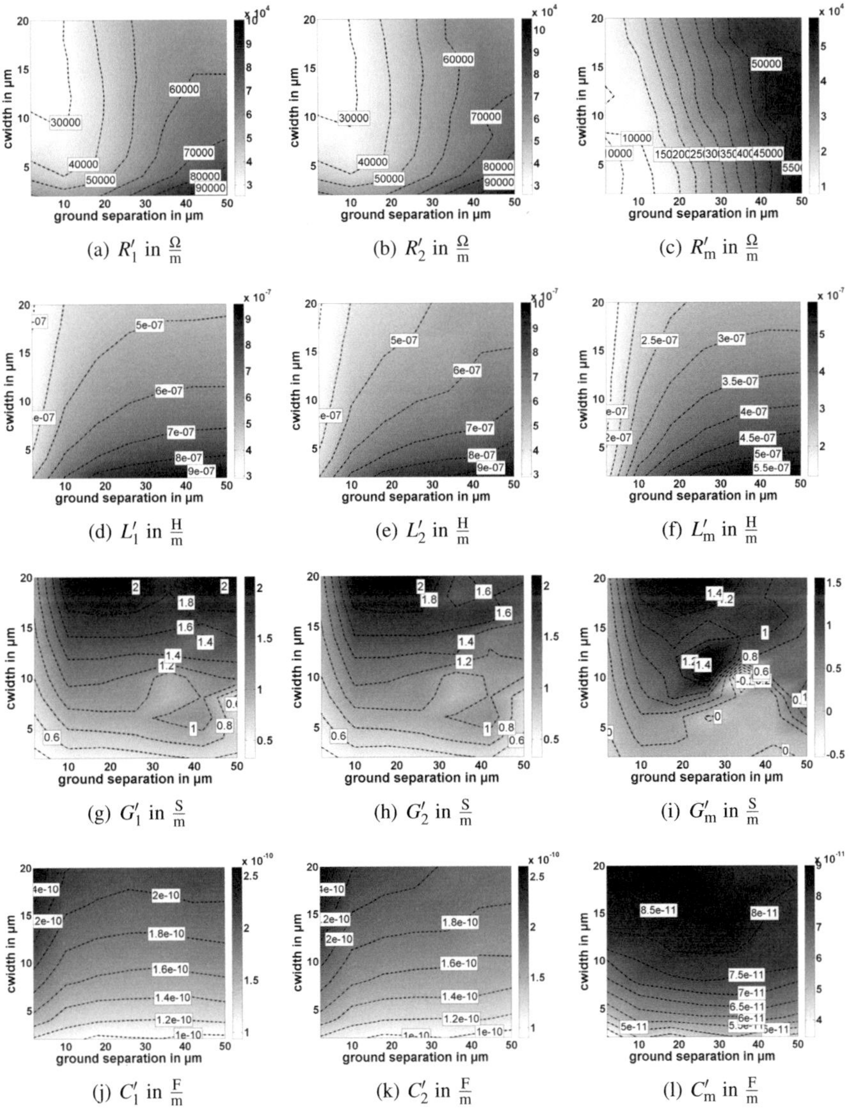

(a) R'_1 in $\frac{\Omega}{\mathrm{m}}$ (b) R'_2 in $\frac{\Omega}{\mathrm{m}}$ (c) R'_m in $\frac{\Omega}{\mathrm{m}}$

(d) L'_1 in $\frac{\mathrm{H}}{\mathrm{m}}$ (e) L'_2 in $\frac{\mathrm{H}}{\mathrm{m}}$ (f) L'_m in $\frac{\mathrm{H}}{\mathrm{m}}$

(g) G'_1 in $\frac{\mathrm{S}}{\mathrm{m}}$ (h) G'_2 in $\frac{\mathrm{S}}{\mathrm{m}}$ (i) G'_m in $\frac{\mathrm{S}}{\mathrm{m}}$

(j) C'_1 in $\frac{\mathrm{F}}{\mathrm{m}}$ (k) C'_2 in $\frac{\mathrm{F}}{\mathrm{m}}$ (l) C'_m in $\frac{\mathrm{F}}{\mathrm{m}}$

Figure F.3: Line parameters of the symmetric coupled line build only with Met1 (configuration a) at 240 GHz. (sep = 2 µm)

F.1.2 Symmetrical coupled line (configuration b)

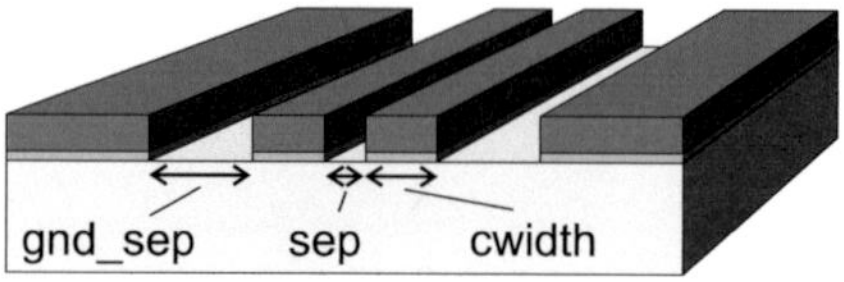

Figure F.4: Geometry of the symmetric coupled line build with Met1 and MetG (configuration b) and realized on IAF's 35 nm and 50 nm process.

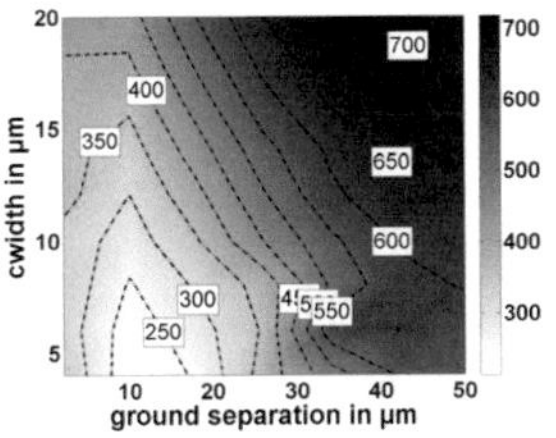

(a) α_c in $\frac{1}{m}$

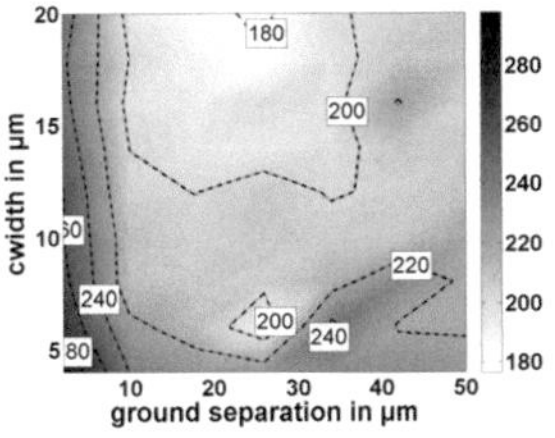

(b) α_π in $\frac{1}{m}$

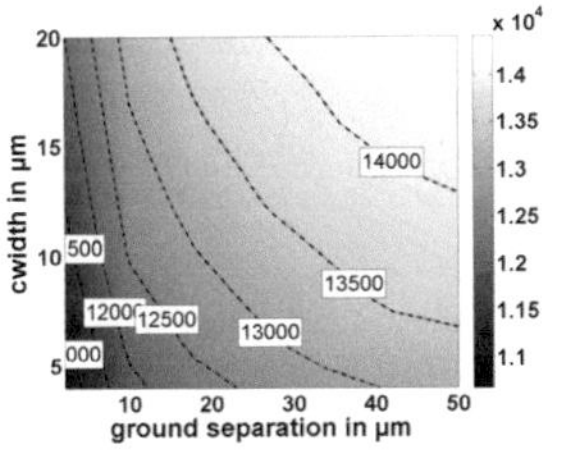

(c) β_c in $\frac{1}{m}$

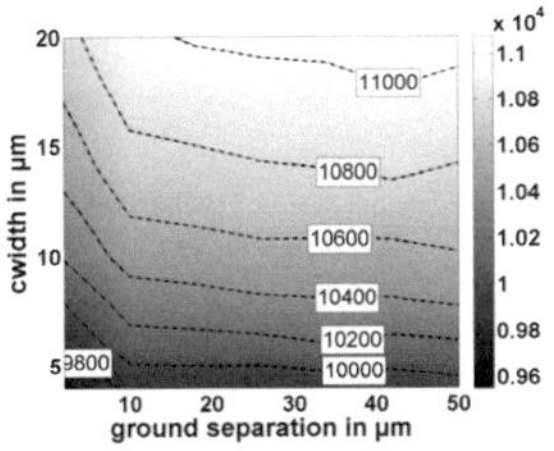

(d) β_π in $\frac{1}{m}$

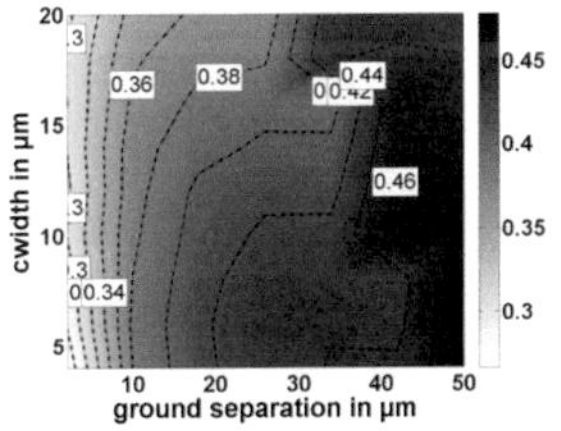

(e) k_c

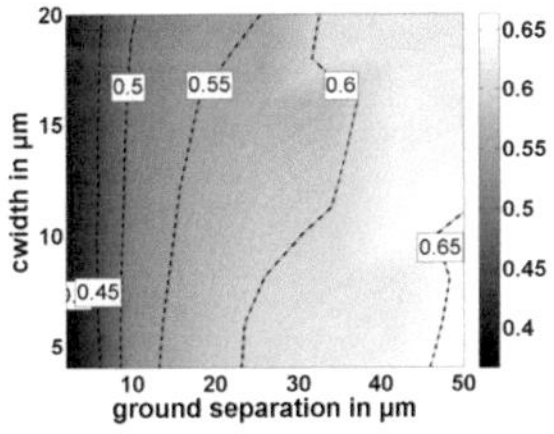

(f) k_m

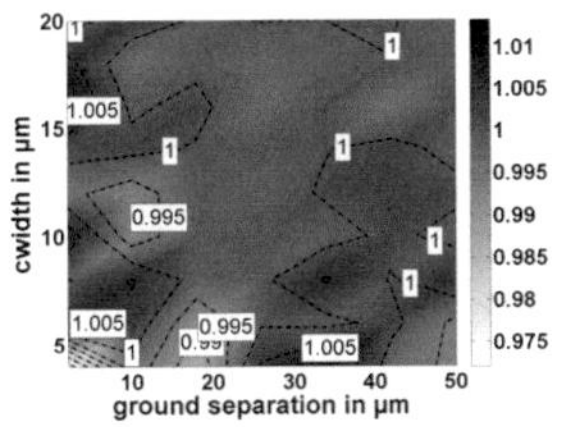

(g) R_c

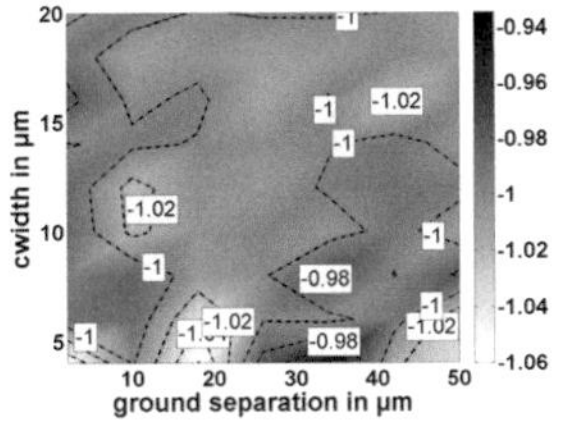

(h) R_π

Figure F.5: Line parameters of the symmetric coupled line build with Met1 and MetG (configuration b) at 240 GHz. (sep = 4 µm)

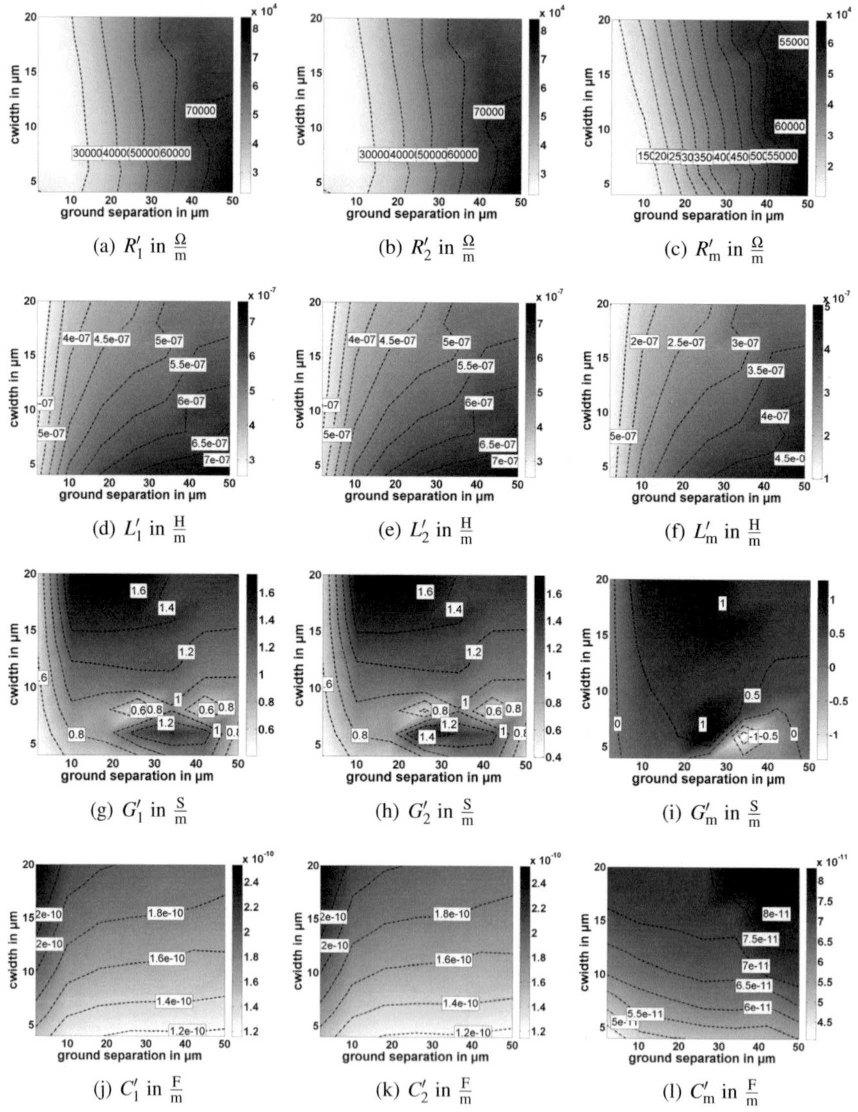

(a) R'_1 in $\frac{\Omega}{\text{m}}$ (b) R'_2 in $\frac{\Omega}{\text{m}}$ (c) R'_m in $\frac{\Omega}{\text{m}}$

(d) L'_1 in $\frac{\text{H}}{\text{m}}$ (e) L'_2 in $\frac{\text{H}}{\text{m}}$ (f) L'_m in $\frac{\text{H}}{\text{m}}$

(g) G'_1 in $\frac{\text{S}}{\text{m}}$ (h) G'_2 in $\frac{\text{S}}{\text{m}}$ (i) G'_m in $\frac{\text{S}}{\text{m}}$

(j) C'_1 in $\frac{\text{F}}{\text{m}}$ (k) C'_2 in $\frac{\text{F}}{\text{m}}$ (l) C'_m in $\frac{\text{F}}{\text{m}}$

Figure F.6: Line parameters of the symmetric coupled line build with Met1 and MetG (configuration b) at 240 GHz. (sep = 4 µm)

F.1.3 Asymmetrical coupled line (configuration c)

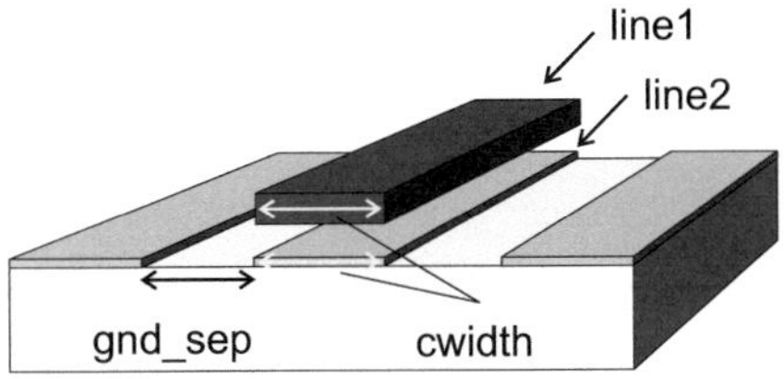

Figure F.7: Geometry of the asymmetric coupled line (configuration c) realized on IAF's 35 nm and 50 nm process.

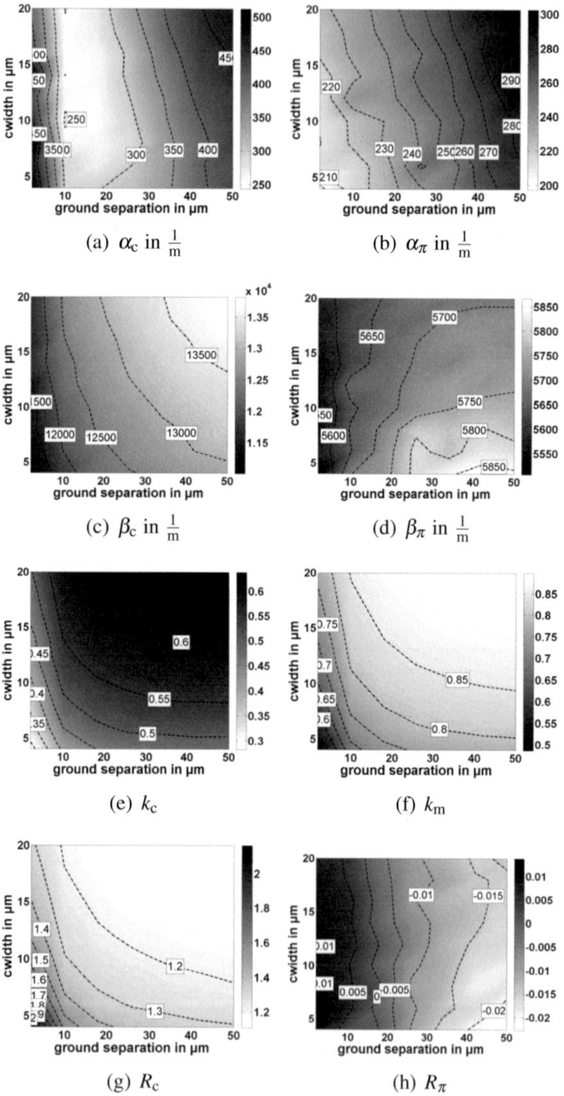

Figure F.8: Line parameters of a asymmetrical coupled line (configuration c) at 240 GHz.

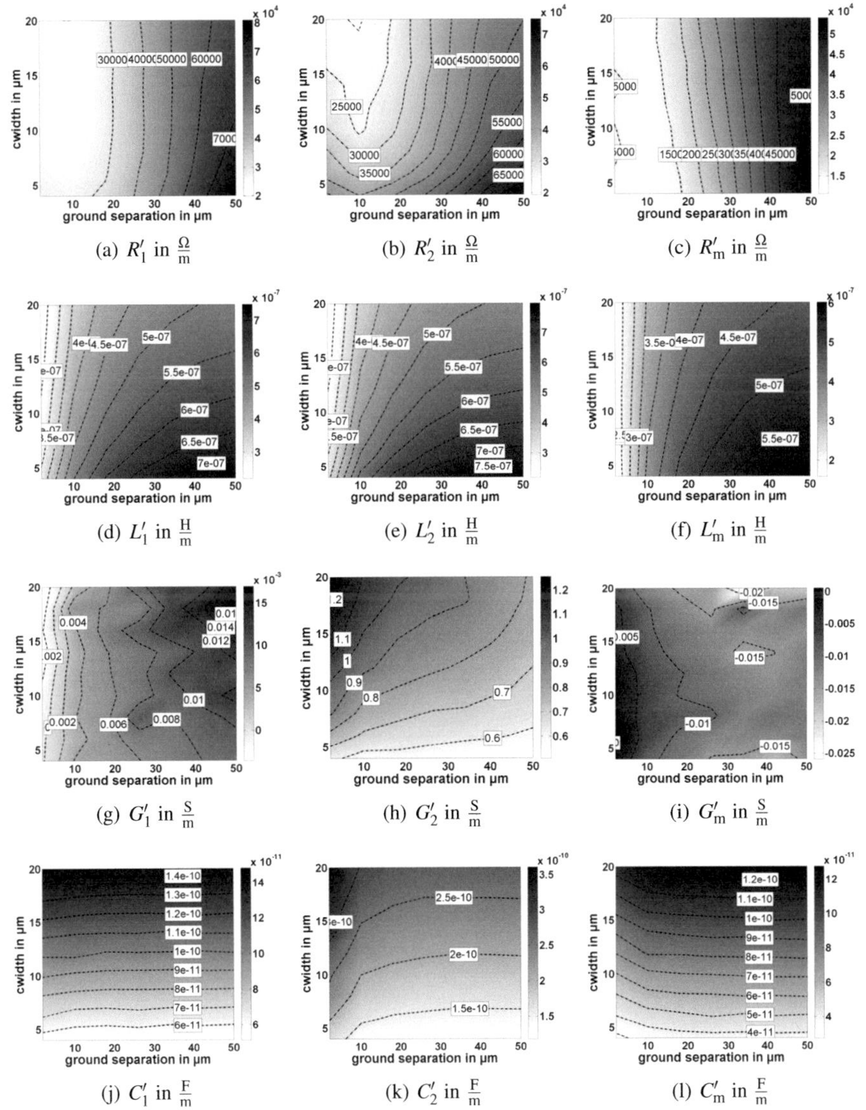

(a) R'_1 in $\frac{\Omega}{\mathrm{m}}$ (b) R'_2 in $\frac{\Omega}{\mathrm{m}}$ (c) R'_m in $\frac{\Omega}{\mathrm{m}}$

(d) L'_1 in $\frac{\mathrm{H}}{\mathrm{m}}$ (e) L'_2 in $\frac{\mathrm{H}}{\mathrm{m}}$ (f) L'_m in $\frac{\mathrm{H}}{\mathrm{m}}$

(g) G'_1 in $\frac{\mathrm{S}}{\mathrm{m}}$ (h) G'_2 in $\frac{\mathrm{S}}{\mathrm{m}}$ (i) G'_m in $\frac{\mathrm{S}}{\mathrm{m}}$

(j) C'_1 in $\frac{\mathrm{F}}{\mathrm{m}}$ (k) C'_2 in $\frac{\mathrm{F}}{\mathrm{m}}$ (l) C'_m in $\frac{\mathrm{F}}{\mathrm{m}}$

Figure F.9: Line parameters of a asymmetric coupled line (configuration c) at 240 GHz.

Bibliography

[ABM$^+$14] J. Antes, F. Boes, D. Meier, T. Messinger, U. Lewark, A. Tessmann, S. Wagner, and I. Kallfass. Ultra-Wideband Single-Balanced Transmitter-MMIC for 300 GHz Communication Systems. In *IEEE MTT-S International Microwave Symposium Digest (IMS)*, June 2014.

[AKLD$^+$13] J. Antes, S. Koenig, D. Lopez-Diaz, F. Boes, A. Tessmann, R. Henneberger, O. Ambacher, T. Zwick, and I. Kallfass. Transmission of an 8-PSK modulated 30 Gbit/s signal using an MMIC-based 240 GHz wireless link. In *IEEE MTT-S International Microwave Symposium Digest (IMS)*, pages 1–3, June 2013.

[AKRH01] I. Aoki, S.D. Kee, D. Rutledge, and A. Hajimiri. A 2.4-GHz, 2.2-W, 2-V fully-integrated CMOS circular-geometry active-transformer power amplifier. In *IEEE Conference on Custom Integrated Circuits Digest*, pages 57–60, 2001.

[AKRH02a] I. Aoki, S.D. Kee, D.B. Rutledge, and A. Hajimiri. Distributed active transformer-A new power-combining and impedance-transformation technique. *IEEE Transactions on Microwave Theory and Techniques*, 50(1):316–331, 2002.

[AKRH02b] I. Aoki, S.D. Kee, D.B. Rutledge, and A. Hajimiri. Fully integrated CMOS power amplifier design using the distributed active-transformer architecture. *IEEE Journal of Solid-State Circuits*, 37(3):371–383, Mar 2002.

[ALD$^+$12] S. Aloui, Y. Luque, N. Demirel, B. Leite, R. Plana, D. Belot, and E. Kerherve. Optimized power combining technique to design a

20dB gain, 13.5dBm OCP1 60GHz power amplifier using 65nm CMOS technology. In *IEEE Radio Frequency Integrated Circuits Symposium (RFIC)*, pages 53–56, June 2012.

[ALK+08] Kyu Hwan An, Ockgoo Lee, Hyungwook Kim, Dong Ho Lee, Jeonghu Han, Ki Seok Yang, Younsuk Kim, Jae Joon Chang, Wangmyong Woo, Chang-Ho Lee, Haksun Kim, and J. Laskar. Power-Combining Transformer Techniques for Fully-Integrated CMOS Power Amplifiers. *IEEE Journal of Solid-State Circuits*, 43(5):1064–1075, May 2008.

[BBC+11] A. Brown, K. Brown, J. Chen, K. C. Hwang, N. Kolias, and R. Scott. W-band GaN power amplifier MMICs. In *IEEE MTT-S International Microwave Symposium Digest (MTT)*, pages 1–4, June 2011.

[BPJ+13] C. Bredendiek, N. Pohl, T. Jaeschke, K. Aufinger, and A. Bilgic. A 240 GHz single-chip radar transceiver in a SiGe bipolar technology with on-chip antennas and ultra-wide tuning range. In *IEEE Radio Frequency Integrated Circuits Symposium (RFIC)*, pages 309–312, June 2013.

[BPKL10] J. Brinkhoff, Duy-Dong Pham, Kai Kang, and F. Lin. A New Six-Port Transformer Modeling Methodology Applied to 10-dBm 60-GHz CMOS ASK Modulator Designs. *IEEE Transactions on Microwave Theory and Techniques*, 58(2):297–309, Feb 2010.

[BSMM13] Il'ja N. Bronštejn, Konstantin A. Semendjaev, Gerhard Musiol, and Heiner Mühlig, editors. *Taschenbuch der Mathematik*. Europa Lehrmittel, Haan-Gruiten, 9. edition, 2013.

[CdNS+07] I. Cendoya, J. de No, B. Sedano, A. Garcia-Alonso, D. Valderas, and I. Gutierrez. A New Methodology for the On-Wafer Characterization of RF Integrated Transformers. *IEEE Transactions on Microwave Theory and Techniques*, 55(5):1046–1053, May 2007.

[CHH+10] C. H Chen, C. H Huang, T. S Horng, J. Y Li, C. C Cheng, C. C Wang, C. T Chiu, and C. P Hung. Generalized efficiency measure for an Above-IC multifilar transformer applied to multiport integrated passive devices. In *60th Electronic Components and Technology Conference Proceedings (ECTC)*, pages 1986–1989, June 2010.

[CIL+98] Y.C. Chen, D.L. Ingram, R. Lai, M. Barsky, R. Grunbacher, T. Block, H.C. Yen, and D.C. Streit. A 95-GHz InP HEMT MMIC amplifier with 427-mW power output. *IEEE Microwave and Guided Wave Letters*, 8(11):399–401, Nov 1998.

[CN11] Jiashu Chen and A.M. Niknejad. A compact 1V 18.6dBm 60GHz power amplifier in 65nm CMOS. In *IEEE International Solid-State Circuits Conference Digest of Technical Papers (ISSCC)*, pages 432–433, Feb 2011.

[Cri06] S.C. Cripps. *RF power amplifiers for wireless communications, second edition*. The Artech House Microwave Library. Artech House, Incorporated, 2006.

[CRN08] D. Chowdhury, P. Reynaert, and A.M. Niknejad. A 60GHz 1V + 12.3dBm Transformer-Coupled Wideband PA in 90nm CMOS. In *IEEE International Solid-State Circuits Conference Digest of Technical Papers (ISSCC)*, pages 560–635, Feb 2008.

[CRN09] D. Chowdhury, P. Reynaert, and A.M. Niknejad. Design Considerations for 60 GHz Transformer-Coupled CMOS Power Amplifiers. *IEEE Journal of Solid-State Circuits*, 44(10):2733–2744, Oct 2009.

[Die14] Sebastian Diebold. *Transistor- und Leitungsmodellierung zum Entwurf von monolithisch integrierten Leistungsverstärkern für den hohen Millimeterwellen-Frequenzbereich*. PhD thesis, KIT, Karlsruhe, 2014.

[DMR+08] W.R. Deal, X.B. Mei, V. Radisic, B. Bayuk, A. Fung, W. Yoshida, P.H. Liu, J. Uyeda, L. Samoska, T. Gaier, and R. Lai. A balanced sub-millimeter wave power amplifier. In *IEEE MTT-S International Microwave Symposium Digest*, pages 399–402, June 2008.

[DPG+12] S. Diebold, P. Pahl, B. Goettel, H. Massler, A. Tessmann, A. Leuther, T. Zwick, and I. Kallfass. A 130 to 160 GHz broadband power amplifier with binary power splitting topology. In *Asia-Pacific Microwave Conference Proceedings (APMC)*, pages 442–444, Dec 2012.

[DWSE+11] S. Diebold, R. Weber, M. Seelmann-Eggebert, H. Massler, A Tessmann, A Leuther, and I Kallfass. A fully-scalable coplanar waveguide passive library for millimeter-wave monolithic integrated circuit design. In *41st European Microwave Conference Proceedings (EuMC)*, pages 293–296, Oct 2011.

[EE92] W.R. Eisenstadt and Y. Eo. S-parameter-based IC interconnect transmission line characterization. *IEEE Transactions on Components, Hybrids, and Manufacturing Technology*, 15(4):483–490, Aug 1992.

[EGKB07] Ouail El-Gharniti, E. Kerherve, and J-B Begueret. Modeling and Characterization of On-Chip Transformers for Silicon RFIC. *IEEE Transactions on Microwave Theory and Techniques*, 55(4):607–615, Apr 2007.

[Ell05] F. Ellinger. 60-GHz SOI CMOS traveling-wave amplifier with NF below 3.8 dB from 0.1 to 40 GHz. *IEEE Journal of Solid-State Circuits*, 40(2):553–558, Feb 2005.

[EMPvR11] J. Essing, R. Mahmoudi, Yu Pei, and A. van Roermund. A fully integrated 60GHz distributed transformer power amplifier in bulky CMOS 45nm. In *IEEE Radio Frequency Integrated Circuits Symposium Digest (RFIC)*, pages 1–4, June 2011.

[Fri46] H.T. Friis. A Note on a Simple Transmission Formula. *Proceedings of the IRE*, 34(5):254–256, May 1946.

[GRRF13] Z. Griffith, T. Reed, M. Rodwell, and M. Field. A 220GHz solid-state power amplifier MMIC with 26.8dB S21 gain, and 55.5mW

Pout at 17.0dB compressed gain. In *IEEE MTT-S International Microwave Symposium Digest (IMS)*, pages 1–3, June 2013.

[GW98] Norbert Geng and Werner Wiesbeck. *Planungsmethoden fuer die Mobilkommunikation: Funknetzplanung unter realen physikalischen Ausbreitungsbedingungen*. Information und Kommunikation. Springer Berlin Heidelberg, 1998.

[GXC12] Q.J. Gu, Z. Xu, and M. C F Chang. Two-Way Current-Combining W -Band Power Amplifier in 65-nm CMOS. *IEEE Transactions on Microwave Theory and Techniques*, 60(5):1365–1374, May 2012.

[HBR+05] Katherine J. Herrick, K.W. Brown, F.A. Rose, Colin S. Whelan, J. Kotce, Jeffrey R. LaRoche, and Yiwen Zhang. W-band metamorphic HEMT with 267 mW output power. In *IEEE MTT-S International Microwave Symposium Digest (IMS)*, pages 843–846, June 2005.

[HCH12] Chien-Hsiang Huang, Chien-Hsun Chen, and Tzyy-Sheng Horng. Novel formulation of efficiency measure for multifilar transformers in generalized multiport configurations of integrated passive devices. *Microwave and Optical Technology Letters*, 54(3):656–661, 2012.

[HCR+08] P. Haldi, D. Chowdhury, P. Reynaert, Gang Liu, and A.M. Niknejad. A 5.8 GHz 1 V Linear Power Amplifier Using a Novel On-Chip Transformer Power Combiner in Standard 90 nm CMOS. *IEEE Journal of Solid-State Circuits*, 43(5):1054–1063, 2008.

[HHKE09] S. Hauptmann, M. Hellfeld, C. Knochenhauer, and F. Ellinger. Modeling bond wires for millimeter wave RFIC design. In *IEEE International Conference on Microwaves, Communications, Antennas and Electronics Systems (COMCAS)*, pages 1–4, Nov 2009.

[HLN06] P. Haldi, G. Liu, and A.M. Niknejad. CMOS compatible transformer power combiner. *IET Electronics Letters*, 42(19):1091–1092, Sept 2006.

[HLR10] Ying He, Lianming Li, and P. Reynaert. 60GHz power amplifier with distributed active transformer and local feedback. In *2010 Proceedings of the ESSCIRC*, pages 314–317, Sept 2010.

[HMM11] J.T. Harvey, S.J. Mahon, and W.F. Montgomery. History and Evolution of Millimetre-Wave MMICs for Point-to-Point Radio. In *IEEE Compound Semiconductor Integrated Circuit Symposium (CSICS)*, pages 1–4, Oct 2011.

[HTL+13] Yuan-Hung Hsiao, Zuo-Min Tsai, Hsin-Chiang Liao, Jui-Chih Kao, and Huei Wang. Millimeter-Wave CMOS Power Amplifiers With High Output Power and Wideband Performances. *IEEE Transactions on Microwave Theory and Techniques*, 61(12):4520–4533, Dec 2013.

[HTLW10] Pin-Cheng Huang, Zuo-Min Tsai, Kun-You Lin, and Huei Wang. A High-Efficiency, Broadband CMOS Power Amplifier for Cognitive Radio Applications. *IEEE Transactions on Microwave Theory and Techniques*, 58(12):3556–3565, Dec 2010.

[HUM+08] Jonathan Hacker, M. Urteaga, Dino Mensa, R. Pierson, Mike Jones, Z. Griffith, and M. Rodwell. 250 nm InP DHBT monolithic amplifiers with 4.8 dB gain at 324 GHz. In *IEEE MTT-S International Microwave Symposium Digest, 2008*, pages 403–406, June 2008.

[HXG+12] Debin Hou, Yong-Zhong Xiong, Wang-Ling Goh, Wei Hong, and M. Madihian. A D-Band Cascode Amplifier With 24.3 dB Gain and 7.7 dBm Output Power in 0.13 um SiGe BiCMOS Technology. *IEEE Microwave and Wireless Components Letters,*, 22(4):191–193, Apr 2012.

[JR05] Sanggeun Jeon and D.B. Rutledge. A 2.7-kW, 29-MHz class-E/Fodd amplifier with a distributed active transformer. In *IEEE MTT-S International Microwave Symposium Digest*, pages 1927–1930, June 2005.

[JTHW09] Yung-Nien Jen, Jeng-Han Tsai, Tian-Wei Huang, and Huei Wang. Design and Analysis of a 55-71-GHz Compact and Broadband Distributed Active Transformer Power Amplifier in 90-nm CMOS Process. *IEEE Transactions on Microwave Theory and Techniques*, 57(7):1637–1646, Jul 2009.

[KAS+11] I. Kallfass, J. Antes, T. Schneider, F. Kurz, D. Lopez-Diaz, S. Diebold, H. Massler, A. Leuther, and A. Tessmann. All Active MMIC-Based Wireless Communication at 220 GHz. *IEEE Transactions on Terahertz Science and Technology*, 1(2):477–487, Nov 2011.

[KBH11] Bon-Hyun Ku, Sang-Hyun Baek, and Songcheol Hong. A Wideband Transformer-Coupled CMOS Power Amplifier for X -Band Multifunction Chips. *IEEE Transactions on Microwave Theory and Techniques*, 59(6):1599–1609, June 2011.

[KGV91] M. C A M Koolen, J. A M Geelen, and M. P J G Versleijen. An improved de-embedding technique for on-wafer high-frequency characterization. In *Proceedings of the 1991 Bipolar Circuits and Technology Meeting*, pages 188–191, Sep 1991.

[KH70] M.K. Krage and G.I. Haddad. Characteristics of Coupled Microstrip Transmission Lines-I: Coupled-Mode Formulation of Inhomogeneous Lines. *IEEE Transactions on Microwave Theory and Techniques*, 18(4):217–222, Apr 1970.

[KH06] A. Komijani and A. Hajimiri. A Wideband 77-GHz, 17.5-dBm Fully Integrated Power Amplifier in Silicon. *IEEE Journal of Solid-State Circuits*, 41(8):1749–1756, Aug 2006.

[KKOY06] Woopoung Kim, Joong-Ho Kim, Dan Oh, and Chuck Yuan. S-parameters Based Transmission Line Modeling with Accurate Low-Frequency Response. In *IEEE Electrical Performance of Electronic Packaging*, pages 79–82, Oct 2006.

[KLL+05] Seungwoo Kim, Kyungho Lee, Jongwoo Lee, Bumman Kim, S.D. Kee, I. Aoki, and D.B. Rutledge. An optimized design of distributed active transformer. *IEEE Transactions on Microwave Theory and Techniques*, 53(1):380–388, Jan 2005.

[KPM+09a] I. Kallfass, P. Pahl, H. Massler, A. Leuther, A. Tessmann, S. Koch, and T. Zwick. A 144 GHz power amplifier MMIC with 11 dBm output power, 10 dB associated gain and 10 % power-added efficiency. In *IEEE MTT-S International Microwave Symposium Digest*, pages 429–432, June 2009.

[KPM+09b] I. Kallfass, P. Pahl, H. Massler, A. Leuther, A. Tessmann, S. Koch, and T. Zwick. A 200 GHz Monolithic Integrated Power Amplifier in Metamorphic HEMT Technology. *IEEE Microwave and Wireless Components Letters*, 19(6):410–412, June 2009.

[LCJ+03] Heeseok Lee, Kiwon Choi, Kyoung-Lae Jang, Taeje Cho, and Seyong Oh. A new efficient equivalent circuit extraction method for multi-port high speed package using multi-input multi-output transmission matrix and polynomial curve fitting. In *53rd Electronic Components and Technology Conference Proceedings*, pages 1582–1588, 2003.

[LDKT+12] D. Lopez-Diaz, I. Kallfass, A. Tessmann, A. Leuther, S. Wagner, M. Schlechtweg, and O. Ambacher. A subharmonic chipset for gigabit communication around 240 GHz. In *IEEE MTT-S International Microwave Symposium Digest (MTT)*, pages 1–3, June 2012.

[LDM+12] J. Langst, S. Diebold, H. Massler, A. Tessmann, A. Leuther, T. Zwick, and I. Kallfass. A balanced 150-240 GHz amplifier MMIC using airbridge transmission lines. In *Workshop on Integrated Nonlinear Microwave and Millimetre-Wave Circuits (INMMIC)*, pages 1–3, Sept 2012.

[LDM+13] J. Langst, S. Diebold, H. Massler, S. Wagner, A. Tessmann, A. Leuther, T. Zwick, and I. Kallfass. Balanced medium power amplifier MMICs from 200 to 270 GHz. In *38th International Conference on Infrared, Millimeter, and Terahertz Waves (IRMMW-THz)*, pages 1–3, Sept 2013.

[Lee04] T.H. Lee. *The Design of CMOS Radio-Frequency Integrated Circuits*. Cambridge University Press, 2004.

[LLN06] Gang Liu, Tsu-Jae King Liu, and A.M. Niknejad. A 1.2V, 2.4GHz Fully Integrated Linear CMOS Power Amplifier with Efficiency Enhancement. In *IEEE Custom Integrated Circuits Conference (CICC)*, pages 141–144, Sept 2006.

[Lon00] J.R. Long. Monolithic transformers for silicon RF IC design. *IEEE Journal of Solid-State Circuits*, 35(9):1368–1382, Sept 2000.

[LPGG09] D. Liu, U. Pfeiffer, J. Grzyb, and B. Gaucher. *Advanced Millimeter-wave Technologies: Antennas, Packaging and Circuits*. Wiley, 2009.

[LPH09] Hongtak Lee, Changkun Park, and Songcheol Hong. A Quasi-Four-Pair Class-E CMOS RF Power Amplifier With an Integrated Passive Device Transformer. *IEEE Transactions on Microwave Theory and Techniques*, 57(4):752–759, Apr 2009.

[LTD+13] A. Leuther, A. Tessmann, M. Dammann, H. Massler, M. Schlechtweg, and O. Ambacher. 35 nm mHEMT Technology for THz and ultra low noise applications. In *International Conference on Indium Phosphide and Related Materials (IPRM)*, pages 1–2, May 2013.

[LTK+10] A. Leuther, A. Tessmann, I. Kallfass, H. Massler, R. Loesch, M. Schlechtweg, M. Mikulla, and O. Ambacher. Metamorphic HEMT technology for submillimeter-wave MMIC applications. In *International Conference on Indium Phosphide Related Materials (IPRM)*, pages 1–6, May 2010.

[LTLW07] Ming-Fong Lei, Zuo-Min Tsai, Kun-You Lin, and Huei Wang. Design and Analysis of Stacked Power Amplifier in Series-Input and Series-Output Configuration. *IEEE Transactions on Microwave Theory and Techniques*, 55(12):2802–2812, Dec 2007.

[LTM+12] A. Leuther, A. Tessmann, H. Massler, R. Aidam, M. Schlechtweg, and O. Ambacher. 450 GHz amplifier MMIC in 50 nm metamorphic HEMT technology. In *International Conference on Indium Phosphide and Related Materials (IPRM)*, pages 229–232, Aug 2012.

[LVG10] Jie-Wei Lai and A. Valdes-Garcia. A 1V 17.9dBm 60GHz power amplifier in standard 65nm CMOS. In *IEEE International Solid-State Circuits Conference Digest of Technical Papers (ISSCC)*, pages 424–425, Feb 2010.

[LW00] T.H. Lee and S.S. Wong. CMOS RF integrated circuits at 5 GHz and beyond. *Proceedings of the IEEE*, 88(10):1560–1571, Oct 2000.

[MBBH07] R.K. Mongia, I.J. Bahl, P. Bhartia, and S.J. Hong. *RF and Microwave Coupled-Line Circuits*. Artech House microwave library. Artech House, Incorporated, 2007.

[MKM+08] M. Micovic, A. Kurdoghlian, H.P. Moyer, P. Hashimoto, M. Hu, M. Antcliffe, P.J. Willadsen, W.-S. Wong, R. Bowen, I. Milosavljevic, Y. Yoon, A. Schmitz, M. Wetzel, C. McGuire, B. Hughes, and D.H. Chow. GaN MMIC PAs for E-Band (71 GHz - 95 GHz) Radio. In *IEEE Compound Semiconductor Integrated Circuits Symposium (CSIC)*, pages 1–4, Oct 2008.

[MKM+12] A. Margomenos, A. Kurdoghlian, M. Micovic, K. Shinohara, D.F. Brown, R. Bowen, I. Milosavljevic, R. Grabar, C. Butler, A. Schmitz, P.J. Willadsen, M. Madhav, and D.H. Chow. 70-105 GHz wideband GaN power amplifiers. In *7th European Microwave Integrated Circuits Conference Digest (EuMIC)*, pages 199–202, Oct 2012.

[MKS+10] B. Martineau, V. Knopik, A. Siligaris, F. Gianesello, and D. Belot. A 53-to-68GHz 18dBm power amplifier with an 8-way combiner in standard 65nm CMOS. In *2010 IEEE International Solid-State Circuits Conference Digest of Technical Papers (ISSCC)*, pages 428–429, Feb 2010.

[MS00] T.K. Moon and W.C. Stirling. *Mathematical Methods and Algorithms for Signal Processing*. Prentice Hall, 2000.

[Nik07] Ali M. Niknejad. *Electromagnetics for High-Speed Analog and Digital Communication Circuits*. Cambridge University Press, 2007.

[NMM+10] Y. Nakasha, Satoshi Masuda, K. Makiyama, T. Ohki, M. Kanamura, Naoya Okamoto, T. Tajima, T. Seino, Hisao Shigematsu, Kenji Imanishi, T. Kikkawa, K. Joshin, and N. Hara. E-Band 85-mW Oscillator and 1.3-W Amplifier ICs Using 0.12um GaN HEMTs for Millimeter-Wave Transceivers. In *IEEE Compound Semiconductor Integrated Circuit Symposium Digest (CSICS)*, pages 1–4, Oct 2010.

[NRB02] Kiat T. Ng, Behzad Rejaei, and J.N. Burghartz. Substrate effects in monolithic RF transformers on silicon. *IEEE Transactions on Microwave Theory and Techniques*, 50(1):377–383, Jan 2002.

[OKH13] Juntaek Oh, Bonhyun Ku, and Songcheol Hong. A 77-GHz CMOS Power Amplifier With a Parallel Power Combiner Based on Transmission-Line Transformer. *IEEE Transactions on Microwave Theory and Techniques*, 61(7):2662–2669, Jul 2013.

[PDK+14] P. Pahl, S. Diebold, S. Krause, H. Gulan, M. Pauli, H. Massler, A. Leuther, I. Kallfass, and T. Zwick. Design and Evaluation of Realizable and Compact Low-Impedance Transmission Lines for Two Top-Metal-Layer Semiconductor Processes. In *Asia-Pacific Microwave Conference Proceedings (APMC)*, Nov 2014.

[PDS+13] P. Pahl, S. Diebold, D. Schwantuschke, S. Wagner, R. Lozar, R. Quay, I. Kallfass, and T. Zwick. A 65-100 GHz impedance transforming hybrid coupler for a V- /W-band AlGaN/GaN MMIC. In *European Microwave Integrated Circuits Conference Proceedings (EuMIC)*, pages 412–415, Oct 2013.

[PG07] U.R. Pfeiffer and D. Goren. A 23-dBm 60-GHz Distributed Active Transformer in a Silicon Process Technology. *IEEE Transactions on Microwave Theory and Techniques*, 55(5):857–865, May 2007.

[PGFR05] U.R. Pfeiffer, D. Goren, B.A. Floyd, and S.K. Reynolds. SiGe transformer matched power amplifier for operation at millimeter-wave frequencies. In *Proceedings of the 31st European Solid-State Circuits Conference (ESSCIRC)*, pages 141–144, Sept 2005.

[PHKH08] Changkun Park, Jeonghu Han, Haksun Kim, and Songcheol Hong. A 1.8-GHz CMOS Power Amplifier Using a Dual-Primary Transformer With Improved Efficiency in the Low Power Region. *IEEE Transactions on Microwave Theory and Techniques*, 56(4):782–792, Apr 2008.

[PJA12] N. Pohl, T. Jaeschke, and K. Aufinger. An Ultra-Wideband 80 GHz FMCW Radar System Using a SiGe Bipolar Transceiver Chip Stabilized by a Fractional-N PLL Synthesizer. *IEEE Transactions on Microwave Theory and Techniques*, 60(3):757–765, Mar 2012.

[PKKH07] Changkun Park, Y. Kim, Haksun Kim, and Songcheol Hong. A 1.9-GHz CMOS Power Amplifier Using Three-Port Asymmetric Transmission Line Transformer for a Polar Transmitter. *IEEE Transactions on Microwave Theory and Techniques*, 55(2):230–238, Feb 2007.

[PMEvR12] Yu Pei, R. Mahmoudi, J. Essing, and A. van Roermund. A 60GHz fully integrated power amplifier using a distributed ring transformer in CMOS 65nm. In *IEEE 19th Symposium on Communications and Vehicular Technology in the Benelux (SCVT)*, pages 1–4, Nov 2012.

[Poz04] D.M. Pozar. *Microwave Engineering*. Wiley, 2004.

[QTK+11] R. Quay, A. Tessmann, R. Kiefer, S. Maroldt, C. Haupt, U. Nowotny, R. Weber, H. Massler, D. Schwantuschke, M. Seelmann-Eggebert, A. Leuther, M. Mikulla, and O. Ambacher. Dual-Gate GaN MMICs for MM-Wave Operation. *IEEE Microwave and Wireless Components Letters*, 21(2):95–97, Feb 2011.

[RDL+10] V. Radisic, W.R. Deal, K.M.K.H. Leong, X.B. Mei, W. Yoshida, Po-Hsin Liu, J. Uyeda, A. Fung, L. Samoska, T. Gaier, and R. Lai. A 10-mW Submillimeter-Wave Solid-State Power-Amplifier Module. *IEEE Transactions on Microwave Theory and Techniques*, 58(7):1903–1909, Jul 2010.

[Rei13] Markus Reinhardt. Optimierung von Kopplerstrukturen aus Basis von elektromagnetischen Feldsimulationen. Studienarbeit, IHE - KIT, 2013.

[RGR+13] T.B. Reed, Z. Griffith, P. Rowell, M. Field, and M. Rodwell. A 180mW InP HBT Power Amplifier MMIC at 214 GHz. In *IEEE Compound Semiconductor Integrated Circuit Symposium Digest (CSICS)*, pages 1–4, Oct 2013.

[RHF12] H. Rucker, B. Heinemann, and A. Fox. Half-Terahertz SiGe BiCMOS technology. In *IEEE 12th Topical Meeting on Silicon Monolithic Integrated Circuits in RF Systems Digest (SiRF)*, pages 133–136, Jan 2012.

[RLM+10] V. Radisic, K.M.K.H. Leong, X. Mei, S. Sarkozy, W. Yoshida, Po-Hsin Liu, J. Uyeda, R. Lai, and W.R. Deal. A 50 mW 220 GHz power amplifier module. In *IEEE MTT-S International Microwave Symposium Digest (MTT)*, pages 45–48, May 2010.

[RRG+11] Thomas B. Reed, Mark J W Rodwell, Z. Griffith, P. Rowell, M. Urteaga, M. Field, and J. Hacker. 48.8 mW Multi-Cell InP HBT Amplifier with On-Wafer Power Combining at 220 GHz. In *IEEE Compound Semiconductor Integrated Circuit Symposium Digest (CSICS)*, pages 1–4, Oct 2011.

[RRG+12a] T.B. Reed, M. Rodwell, Z. Griffith, P. Rowell, A. Young, M. Urteaga, and M. Field. A 220 GHz InP HBT Solid-State Power Amplifier MMIC with 90mW POUT at 8.2dB Compressed Gain. In *IEEE Compound Semiconductor Integrated Circuit Symposium Digest (CSICS)*, pages 1–4, Oct 2012.

[RRG+12b] Thomas B. Reed, Mark J W Rodwell, Z. Griffith, P. Rowell, M. Field, and M. Urteaga. A 58.4mW solid-state power amplifier at 220 GHz using InP HBTs. In *IEEE MTT-S International Microwave Symposium Digest (MTT)*, pages 1–3, June 2012.

[Sam08] M.K. Sampath. On addressing the practical issues in the extraction of RLGC parameters for lossy multiconductor transmission lines using S-parameter models. In *IEEE-EPEP Electrical Performance of Electronic Packaging*, pages 259–262, Oct 2008.

[SCP13] N. Sarmah, P. Chevalier, and U.R. Pfeiffer. 160-GHz Power Amplifier Design in Advanced SiGe HBT Technologies With P_{sat} in Excess of 10 dBm. *IEEE Transactions on Microwave Theory and Techniques*, 61(2):939–947, Feb 2013.

[Sha49] C.E. Shannon. Communication in the Presence of Noise. *Proceedings of the IRE*, 37(1):10–21, Jan 1949.

[SHP13] N. Sarmah, B. Heinemann, and U.R. Pfeiffer. A 135-170 GHz power amplifier in an advanced sige HBT technology. In *IEEE Radio Frequency Integrated Circuits Symposium Digest (RFIC)*, pages 287–290, June 2013.

[Sko03] Skolnik. *Introduction to Radar Systems*. Tata McGraw Hill, 2003.

[SL01] L. Samoska and Yoke Choy Leong. 65-145 GHz InP MMIC HEMT medium power amplifiers. In *IEEE MTT-S International Microwave Symposium Digest*, volume 3, pages 1805–1808, May 2001.

[SL12] S.M. Sze and M.K. Lee. *Semiconductor Devices: Physics and Technology*. Wiley, 2012.

[SPH+04] L. Samoska, A. Peralta, Ming Hu, M. Micovic, and A. Schmitz. A 20 mW, 150 GHz InP HEMT MMIC power amplifier module. *IEEE Microwave and Wireless Components Letters*, 14(2):56–58, Feb 2004.

[STK+11] M. Schlechtweg, A. Tessmann, I. Kallfass, A. Leuther, V. Hurm, H. Massler, M. Riessle, R. Losch, and O. Ambacher. Millimeter-wave circuits and modules up to 500 GHz based on metamorphic HEMT technology for remote sensing and wireless communication applications. In *IEEE 9th International New Circuits and Systems Conference (NEWCAS)*, pages 269–272, June 2011.

[SUC12] SUCCESS. *Press release: Measureing Distance and Speed with a thumbtack-sized Radar Sensor*, September 2012. http://www.success-project.eu.

[SYK+07] Jinglin Shi, Wen-Yan Yin, Kai Kang, Jun-Fa Mao, and Le-Wei Li. Frequency-Thermal Characterization of On-Chip Transformers With Patterned Ground Shields. *IEEE Transactions on Microwave Theory and Techniques*, 55(1):1–12, Jan 2007.

[TH03] L.F. Tiemeijer and R.J. Havens. A calibrated lumped-element de-embedding technique for on-wafer RF characterization of high-quality inductors and high-speed transistors. *IEEE Transactions on Electron Devices*, 50(3):822–829, Mar 2003.

[TLM+08] A. Tessmann, A. Leuther, H. Massler, M. Kuri, and R. Loesch. A Metamorphic 220-320 GHz HEMT Amplifier MMIC. In *Compound Semiconductor Integrated Circuits Symposium Digest (CSIC)*, pages 1–4, Oct 2008.

[TLSM05] A. Tessmann, A. Leuther, C. Schwoerer, and H. Massler. Metamorphic 94 GHz power amplifier MMICs. In *IEEE MTT-S International Microwave Symposium Digest*, pages 1579–1582, June 2005.

[Tri75] V.K. Tripathi. Asymmetric Coupled Transmission Lines in an Inhomogeneous Medium. *IEEE Transactions on Microwave Theory and Techniques*, 23(9):734–739, 1975.

[TTB+13] M. Thian, M. Tiebout, N.B. Buchanan, V.F. Fusco, and F. Dielacher. A 76-84 GHz SiGe Power Amplifier Array Employing Low-Loss Four-Way Differential Combining Transformer. *IEEE Transactions on Microwave Theory and Techniques*, 61(2):931–938, Feb 2013.

[TWK98] Manfred K. A. Thumm, Werner Wiesbeck, and Stefan Kern. *Hochfrequenzmeßtechnik : Verfahren und Meßsysteme*. Teubner, Stuttgart, 2 edition, 1998.

[VVB04] B.S. Virdee, A.S. Virdee, and B.Y. Banyamin. *Broadband Microwave Amplifiers*. Artech House microwave library. Artech House, 2004.

[WCW12] Kun-Yin Wang, Tao-Yao Chang, and Chorng-Kuang Wang. A 1V 19.3dBm 79GHz power amplifier in 65nm CMOS. In *IEEE International Solid-State Circuits Conference Digest of Technical Papers (ISSCC)*, pages 260–262, Feb 2012.

[Wei03] T. Weiland. RF microwave simulators - from component to system design. In *33rd European Microwave Conference Proceedings*, volume 2, pages 591–596, Oct 2003.

[Wir10] WirelessHD. *WirelessHD Specification Version 1.1 Overview*, May 2010. http://www.wirelesshd.org/pdfs/WirelessHD-Specification-Overview-v1.1May2010.pdf.

[WLX+09] Chuan Wang, Huailin Liao, Yongzhong Xiong, Chen Li, Ru Huang, and Yangyuan Wang. A Physics-Based Equivalent-Circuit Model for On-Chip Symmetric Transformers With Accurate Substrate Modeling. *IEEE Transactions on Microwave Theory and Techniques*, 57(4):980–990, Apr 2009.

[WM92] D.F. Williams and Roger B. Marks. Calibrating on-wafer probes to the probe tips. In *40th ARFTG Conference Digest*, volume 22, pages 136–143, Dec 1992.

[YXVG11] S.K. Yong, P. Xia, and A. Valdes-Garcia. *60GHz Technology for Gbps WLAN and WPAN: From Theory to Practice*. Wiley, 2011.

[Zwi05] Thomas Zwick. How-To: Calibrating and De-Embedding. Personal manual - circulated within IBM (Thomas J. Watson Research Center - Yorktown Heights, NY, US), Apr 2005.